AF560516

Control and Analysis for Food and Agricultural Products

Control and Analysis for Food and Agricultural Products

Barun Guha

RANDOM PUBLICATIONS

NEW DELHI - 110 002 (INDIA)

Control and Analysis for Food and Agricultural Products

ISBN 978-93-51112-76-1

Published in 2014 in India by
RANDOM PUBLICATIONS
4376-A/4B, Gali Murari Lal, Ansari Road
New Delhi-110 002
Phone: +9111-43580356, 23289044
E-mail: randomexports@gmail.com; sales@randompublications.com; info@randompublications.com

Reprinted 2021

Type Setting by: Friends Media, Delhi-110089
Digitally Printed at : Replika Press Pvt. Ltd.

Preface

Food analysis is the discipline dealing with the development, application and study of analytical procedures for characterizing the properties of foods and their constituents. These analytical procedures are used to provide information about a wide variety of different characteristics of foods, including their composition, structure, physicochemical properties and sensory attributes. This information is critical to our rational understanding of the factors that determine the properties of foods, as well as to our ability to economically produce foods that are consistently safe, nutritious and desirable and for consumers to make informed choices about their diet. The objective of this book is to review the basic principles of the analytical procedures commonly used to analyze foods and to discuss their application to specific food components, *e.g.* lipids, proteins, water, carbohydrates and minerals. The following questions will be addressed in this introductory section: Who analyzes foods? Why do they analyze foods? What types of properties are measured? How does one choose an appropriate analytical technique for a particular food? The price of certain foods is dictated by the quality of the ingredients that they contain. For example, a packet of premium coffee may claim that the coffee beans are from Columbia, or the label of an expensive wine may claim that it was produced in a certain region, using a certain type of grapes in a particular year. How do we verify these claims? There are many instances in the past where manufacturers have made false claims about the authenticity of their products in order to get a higher price. It is therefore important to have analytical techniques that can be used to test the authenticity of certain food components, to ensure that consumers are not the victims of economic fraud and that competition among food manufacturers is fair.

The government has a *Food Inspection and Grading Service* that routinely analyses the properties of food products to ensure that they meet the appropriate laws and regulations. Hence, both government agencies and food manufacturers need analytical techniques to provide the appropriate information about food properties. The most important criteria for this type of test are often the accuracy of the measurements and the use of an official method. The government has recently carried out a survey of many of the official analytical techniques developed to analyze foods, and has specified which techniques must be used to analyze certain food components for labelling purposes. Techniques have been chosen which provide accurate and reliable results, but which are relatively simple and inexpensive to perform. The food industry is highly competitive and food manufacturers are continually trying to increase their market-share and profits. To do this they must ensure that their products are of *higher quality*, *less expensive*, and more *desirable* than their competitors, whilst ensuring that they are *safe* and *nutritious*. To meet these rigorous standards food manufacturers need analytical techniques to analyze food materials before, during and after the manufacturing process to ensure that the final product meets the desired standards. In a food factory one starts with a number of different raw materials, processes them in a certain manner (*e.g.* heat, cool, mix, dry), packages them for consumption and then stores them. The food is then transported to a warehouse or retailer where it is sold for consumption. One of the most important concerns of the food manufacturer is to produce a final product that consistently has the same overall properties, *i.e.* appearance, texture, flavour and shelf life. When we purchase a particular food product we expect its properties to be the same (or very similar) to previous times, and not to vary from purchase-to-purchase. Ideally, a food manufacture wants to take the raw ingredients, process them in a certain way and produce a product with specific desirable properties. Unfortunately, the properties of the raw ingredients and the processing conditions vary from time to time which causes the properties of the final product to vary, often in an unpredictable way.

Hope this book will be useful to students as reference book and will be a priced collection for their own library.

I thank all members of my team who have helped in the preparation of the book. My special thanks go to "Random Publications" who have published the book.

—Barun Guha

Contents

1

Food Microbiology Basics

Focus of Food Microbiology

The focus of Food Microbiology is on the detection and analysis of foodborne spoilage microorganisms. Food microbiology is the study of food micro-organisms; how we can identify and culture them, how they live, how some infect and cause disease and how we can make use of their activities. Microbes are single-cell organisms so tiny that millions can fit into the eye of a needle. They are the oldest form of life on earth. Microbe fossils date back more than 3.5 billion years to a time when the Earth was covered with oceans that regularly reached the boiling point, hundreds of millions of years before dinosaurs roamed the earth. The field of food microbiology is a very broad one, encompassing the study of microorganisms which have both beneficial and deleterious effects on the quality and safety of raw and processed foods. Food science is a discipline concerned with all aspects of food-beginning after harvesting, and ending with consumption by the consumer. It is considered one of the agricultural sciences, and it is a field which is entirely distinct from the field of nutrition. In the U.S., food science is typically studied at land-grant universities.

Examples of the activities of food scientists include the development of new food products, design of processes to produce these foods, choice of packaging materials, shelf-life studies, sensory evaluation of the product with potential consumers, microbiological and chemical testing. Food scientists in universities may study more fundamental phenomena that are directly linked to the production of a particular food product.

Food scientists are generally not directly involved with the creation of genetically modified (bio-engineered) foods.

Some of the subdisciplines of food science: Food safety, Food engineering, Product development, Sensory analysis, Food chemistry. The primary tool of microbiologists is the ability to identify and quantitate food-borne microorganisms; however, the inherent inaccuracies in enumeration processess, and the natural variation found in all bacterial populations complicate the microbiologists job. Without microbes, we couldn't eat or breathe.

Without us, they'd probably be just fine.

Understanding microbes is vital to understanding the past and the future of ourselves and our planet.

Archaea look and act a lot like bacteria. So much so that until the late 1970s, scientists assumed they were a kind of "weird" bacteria.

Then microbiologist Carl Woese devised an ingenious method of comparing genetic information showing that they could not rightly be called bacteria at all. Their genetic recipe is too different.

So different Woese decided they deserved their own special branch on the great family tree of life, a branch he dubbed the Archaea.

Archaea comes from the Greek word meaning "ancient." An appropriate name, because many archaea thrive in conditions mimicking those found more than 3.5 billion years ago. Back then, the earth was still covered by oceans that regularly reached the boiling point — an extreme condition not unlike the hydrothermal vents and sulfuric waters where archaea are found today.

Some scientists consider archaea living fossils that may provide hints about what the earliest life forms on Earth were like, and how life evolved on our planet.

The Extremists

In addition to superheated waters, archaea have been found in acid-laden streams around old mines, in frigid Antarctic ice and in the super-salty waters of the Dead Sea. A number of other extreme-living bacterial species also enjoy these conditions, too, such as the community of cyanobacteria and bacteria shown top right.

Foodborne illness or food poisoning is caused by consuming food contaminated with pathogenic bacteria, toxins, viruses, prions or parasites. Such contamination usually arises from improper handling, preparation or storage of food. Foodborne illness can also be caused by adding pesticides or medicines to food, or by accidentally consuming naturally poisonous substances like poisonous mushrooms or reef fish.

Contact between food and pests, especially flies, rodents and cockroaches, is a further cause of contamination of food.

Some common diseases are occasionally foodborne mainly through the water vector, even though they are usually transmitted by other routes. These include infections caused by Shigella, Hepatitis A, and the parasites Giardia lamblia and Cryptosporidium parvum.

- Thermophiles like unusually hot temperatures. A few species have been found to survive even above 110 degrees Celsius (water boils at 100 degrees Celsius).
- Psychrophiles like extremely cold temperatures (even down to-10 degrees Celsius).
- Halophiles thrive in unusually salty habitats. Some can thrive in water that's 9% salt; sea water contains only 0.9% salt.
- Acidophiles prefer acidic conditions; Alkaliphiles prefer very alkaline environs.

Accumulating sufficient data on the behaviour of microorganisms in foods requires an extensive amount of work, and is costly. In addition, while data alone can describe the response of a microorganisms in food, they provides little insight into the relationship between physiological processes and growth or survival.

Scientists also use molecular tools to extract and compare bits of a particular kind of RNA from samples in order to determine if previously known or new microbes are present in a particular environment. This technique is widely used as a biomarker and for microbial ecology studies. It uses a particular kind of RNA known as 16S ribosomal RNA, or 16S rRNA.

Ribosomes are the gene-translating machines in all living things. When a gene on a piece of DNA is copied into a strand of messenger RNA and ferried out of the cell nucleus into the cell fluid, ribosomes there latch onto this mRNA. The ribosomes move along the mRNA strand, reading the code contained in its sequence of nucleotide bases (the As, Gs, Cs and Us, since U replaces T in RNA) and stringing the right amino acids together based on the code to build protein chains.

Pasteurization is typically associated with milk. There are two widely used methods to pasteurize milk: high temperature/short time (HTST), and ultra-high temperature (UHT). HTST is by far the most common method. Milk simply labelled "pasteurized" is usually treated with the HTST method, whereas milk labelled "ultra-pasteurized" must be treated with the UHT method. HTST involves holding the milk at a temperature of 161.5 degrees F (or 72 C) for at least 15 seconds. UHT involves holding the milk at a temperature of 280 degrees F (or 138 C) for at least two seconds.

Figure: *Cream pasteurising and cooling coils at Murgon Butter Factory, 1939*

Pasteurization methods are usually standardized and controlled by national food safety agencies (such as the USDA in the United States and the Food Standards Agency in the U.K.). These agencies require milk to be HTST pasteurized in order to qualify for the "pasteurized" label. There are different standards for different dairy products, depending on the fat content and the intended usage. For example, the pasteurization standards for cream differ from the standards for fluid milk, and the standards for pasteurizing cheese are designed to preserve the phosphatase enzyme, which aids in curing the cheese.

The HTST pasteurization standard was designed to achieve a 5-log (approximately one million-fold) reduction in the number of viable microorganisms in milk. This is considered adequate for destroying almost all yeasts, mold, and common spoilage bacteria and also to ensure adequate destruction of common pathogenic heat-resistant organisms (including particularly Mycobacterium tuberculosis, which causes tuberculosis and Coxiella burnetii, which causes Q fever). HTST pasteurization processes must be designed so that the milk is heated evenly, and no part of the milk is subject to a shorter time or a lower temperature. HTST pasteurized milk typically has a refrigerated shelf life of two to three weeks, whereas ultra pasteurized milk can last much longer when refrigerated, sometimes two to three months. When UHT pasteurization is combined with sterile handling and container technology, it can even be stored unrefrigerated for long periods of time.

Direct detection and identification of lactic acid bacteria in a food processing plant and in meat products using denaturing gradient gel electrophoresis; Takahashi H et al.; We established a novel system using denaturing gradient gel electrophoresis (DGGE) to quickly identify bacteria known to be responsible for spoilage in meat processing plants and meat products. We extracted bacterial DNA from swabbed samples at various locations in the plant and from meat products and performed PCR amplification, targeting 16S rDNA from the dominant organisms. The amplification products were subjected to DGGE, and the contaminating bacteria in the meat products and the plant were analysed. This analysis indicated that lactic acid bacteria and spoilage-causing bacteria are widely distributed within the meat processing plant. We developed molecular size markers to identify the dominant organisms obtained from the plant and meat products. The establishment of the present method allows quick and simple identification of bacteria causing the possible deterioration of products and contamination and thus permits constant monitoring of any harmful bacteria within meat processing plants.

Control of Listeria monocytogenes on frankfurters with antimicrobials in the formulation and by dipping in organic acid solutions; Barmpalia IM et al.; The antilisterial activity of sodium lactate (SL) and sodium diacetate (SD) was evaluated in a frankfurter formulation and in combination with a dipping treatment into solutions of lactic acid or acetic acid after processing and inoculation. Pork frankfurters were formulated with 1.8% SL or 0.25% SD or combinations of 1.8% SL with 0.25 or 0.125% SD. After processing, frankfurters were inoculated (2 to 3 log CFU/cm^2) with a 10-strain composite of Listeria monocytogenes and left undipped or were dipped (2 min) in 2.5% solutions of lactic acid or acetic acid (23 +/-2 degrees C) before vacuum packaging and storage at 10 degrees C for 40 days. Total microbial populations and L. monocytogenes, lactic acid bacteria, and yeasts and molds were enumerated during storage. Sensory evaluations also were carried out on frankfurters treated and/or formulated with effective antimicrobials.

The combination of 1.8% SL with 0.25% SD provided complete inhibition of L. monocytogenes growth throughout storage. Dipping in lactic acid or acetic acid reduced initial populations by 0.7 to 2.1 log CFU/cm2, but during storage (12 to 20 days), populations on dipped samples without antimicrobials in the formulation reached 5.5 to 7.9 log CFU/cm2. For samples containing single antimicrobials and dipped in lactic acid or acetic acid, L. monocytogenes growth was completely inhibited or reduced over 12 and 28 days, respectively, whereas final populations were lower ($P < 0.05$) than those in undipped samples of the same formulations. Bactericidal effects during storage (reductions of 0.6 to 1.0 log CFU/cm2 over 28 to 40 days) were observed in frankfurters containing combinations of SL and SD that were dipped in organic acid solutions. Inclusion of antimicrobials in the formulation and/or dipping the product into organic acid solutions did not affect ($P > 0.05$) the flavour and overall acceptability of products compared with controls. The results of this study may be valuable to meat processors as they seek approaches for meeting new regulatory requirements in the United States.

Comparison of effect of vacuum-condensed and ultrafiltered milk on cheddar cheese; Acharya MR et al.; The objective of this study was to compare the effects of vacuum-condensed (CM) and ultrafiltered (UF) milk on some compositional and functional properties of Cheddar cheese. Five treatments were designed to have 2 levels of concentration (4.5 and 6.0% protein) from vacuum-condensed milk (CM1 and CM2) and ultrafiltered milk (UF1 and UF2) along with a 3.2% protein control. The samples were analysed for fat, protein, ash, calcium, and salt contents at 1 wk. Moisture content, soluble protein, meltability, sodium dodecyl sulfate-PAGE, and counts of lactic acid bacteria and nonstarter lactic acid bacteria were performed on samples at 1, 18, and 30 wk. At 1 wk, the moisture content ranged from 39.2 (control) to 36.5% (UF2).

Fat content ranged from 31.5 to 32.4% with no significant differences among treatments, and salt content ranged from 1.38 to 1.83% with significant differences. Calcium content was higher in UF cheeses than in CM cheeses followed by control, and it increased with protein content in cheese milk. Ultrafiltered milk produced cheese with higher protein content than CM milk. The soluble protein content of all cheeses increased during 30 wk of ripening. Condensed milk cheeses exhibited a higher level of proteolysis than UF cheeses. Sodium dodecyl sulfate-PAGE showed retarded proteolysis with increase in level of concentration. The breakdown of alphas1-casein and alphas1-

I-casein fractions was highest in the control and decreased with increase in protein content of cheese milk, with UF2 being the lowest. There was no significant degradation of beta-casein. Overall increase in proteolytic products was the highest in control, and it decreased with increase in protein content of cheese milk.

No significant differences in the counts of lactic starters or nonstarter lactic acid bacteria were observed. Extent as well as method of concentration influenced the melting characteristics of the cheeses. Melting was greatest in the control cheeses and least in cheese made from condensed milk and decreased with increasing level of milk protein concentration. Vacuum condensing and ultrafiltration resulted in Cheddar cheeses of distinctly different quality. Although both methods have their advantages and disadvantages, the selection of the right method would depend upon the objective of the manufacturer and intended use of the cheese.

Physicochemical and microbiological profile of packed table olives; Lopez-Lopez A et al.; The water activity of different commercial products of table olives as well as their physicochemical and microbiological profiles were studied. Average values of water activity were 0.976, 0.977, and 0.990 for green, directly brined, and ripe (by alkaline oxidation) olives, respectively. Mean values of pH were 3.69, 3.92, and 6.52 while salt levels were 5.53, 4.98, and 2.55 for the same commercial products. In some of the commercial products of green and directly brined olives that were examined, a certain level of microorganisms was observed.

Most of them were lactic acid bacteria and yeasts, the lactic acid bacteria being more abundant than the yeasts. Traditionally, this flora has not been considered harmful due to its resemblance to that found in natural and spontaneous fermentation. Spore-forming mesophilic aerobes, which can be considered contaminants from ingredient or product handling, were present in only some samples and were always at low levels. However, their detection does not cause any concern regarding safety because there is no reference showing that they can grow in any of the products studied given the physicochemical characteristics found.

Biodiversity of Listeria monocytogenes sensitivity to bacteriocin-producing Carnobacterium strains and application in sterile cold-smoked salmon; Brillet A et al.; AIMS: The aim of this study was to demonstrate the inhibitory capacity of Carnobacterium strains against a collection of Listeria monocytogenes strains in cold-smoked salmon (CSS).

Methods and Results

Three bacteriocin-producing strains, Carnobacterium divergens V41, C. piscicola V1 and C. piscicola SF668, were screened for their antilisterial activity against a collection of 57 L. monocytogenes strains selected from the French smoked salmon industry, using an agar spot test. All the Listeria strains were inhibited but three different groups could be distinguished differing in sensitivity to the three Carnobacterium strains.

However, C. divergens V41 always had the highest inhibitory effect. The antilisterial capacity was then tested in sterile CSS blocks co-inoculated with Carnobacterium spp. and mixtures of L. monocytogenes strains. C. divergens V41 was the most efficient strain, maintaining the level of L. monocytogenes at <50 CFU g(-1) during the 4 weeks of vacuum storage at 4 and 8 degrees C, whatever the sensitivity of the set of L. monocytogenes strains. C. divergens V41 may be a good candidate for biopreservation in CSS. A biopreservation strategy for CSS against the risk of L. monocytogenes was investigated using bacteriocin-producing lactic acid bacteria.

Effect of a gamma-aminobutyric acid-enriched dairy product on the blood pressure of spontaneously hypertensive and normotensive Wistar-Kyoto rats; Hayakawa K et al.; We investigated the blood-pressure-lowering effects of gamma-aminobutyric acid (GABA) and a GABA-enriched fermented milk product (FMG) by low-dose oral administration to spontaneously hypertensive (SHR/Izm) and normotensive Wistar-Kyoto (WKY/Izm) rats. FMG was a non-fat fermented milk product produced by lactic acid bacteria, and the GABA contained in FMG was made from the protein of the milk during fermentation. A single oral dose of GABA or FMG (5 ml/kg; 0.5 mg GABA/kg) significantly (P<0.05) decreased the blood pressure of SHR/Izm from 4 to 8 h after administration, but did not increase that of WKY/Izm rats.

The hypotensive activity of GABA was dose-dependent from 0.05 to 5.00 mg/kg in SHR/Izm. During the chronic administration of experimental diets to SHR/Izm, a significantly slower increase in blood pressure with respect to the control group was observed at 1 or 2 weeks after the start of feeding with the GABA or FMG diet respectively (P<0.05) and this difference was maintained throughout the period of feeding. The time profile of blood-pressure change due to administration of FMG was similar to that of GABA. FMG did not inhibit angiotensin 1-converting enzyme. Furthermore, an FMG peptide-containing fraction from reverse-phase chromatography lacked a hypotensive effect in SHR/Izm rats. The present results suggest that low-dose oral GABA has a hypotensive effect in SHR/Izm and that the hypotensive effect of FMG is due to GABA.

The 'buttery' attribute of wine—diacetyl—desirability, spoilage and beyond; Bartowsky EJ et al.; The diketone, diacetyl, is a major flavour metabolite produced by lactic acid bacteria (LAB). Of the LAB associated with wine, Oenococcus oeni is encouraged during the malolactic (ML) fermentation, a biodeacidification of wine during which the metabolism of diacetyl occurs. Diacetyl, which imparts a buttery aroma and flavour to many fermented foods and beverages, is a key flavour compound of most fermented dairy products. In wine, diacetyl has important stylistic implications. The biosynthesis of diacetyl is dependent upon citric acid metabolism and diacetyl is an intermediate metabolite which can be further reduced to acetoin and the alcohol, 2,3-butanediol. This review will focus on the sensory perception, metabolism, genetics and analysis of diacetyl during wine production. The extensive knowledge of diacetyl metabolism in dairy LAB is used to enhance the understanding of diacetyl metabolism of wine LAB. Factors which can effect the formation and concentration of diacetyl in wine are discussed. These include malolactic bacterial strain, wine chemical and physical parameters (pH, temperature, citric acid, sulphur dioxide, aeration) and the presence of yeast lees. Finally, the affects of other wine components, such as phenolics, are discussed.

Objective

To investigate the influence of early enteral nutrition with synbiotics on the plasma endotoxin level, the nutritional state, the inflammatory response and the incidence of infectious complications in severely burned patients.

Methods

Randomized double blind and control method was employed im the study. Forty severely burned patients were randomly divided into A and B groups with 20 in each group. The patients in group A received early enteral nutrition with synbiotics including four kinds of lactic acid bacteria and four kinds of fibres, while those in group B received early enteral nutrition with synbiotics including only four kinds of fibres. The patients with 80% to 280% coefficient unit burned surface (UBS) were further divided into A1 (n = 10) and B1 (n = 11) groups. The plasma endotoxin level in group A and B was determined dynamically on the 1st, 3rd, 7th, 10th, 14th, and 21st postburn days (PBD), and its abnormal rate in both groups was statistically analysed in correlation with the normal endotoxin level. Meanwhile, the mortality, the incidence of infectious complications and the blood bacterial culture results were compared between the two groups. The plasma levels of IL-1, IL-6 and prognostic inflammatory nutrition index (PINI) were also determined at the above time points.

The plasma endotoxin level in group A (37.9 +/-5.4) ng/L was evidently lower than that in group B (59.1 +/-7.9) ng/L ($P < 0.05$) on 10 PBD. The abnormal rate of plasma endotoxin in group A (36.7%) was evidently lower than that (49.2%) in group B ($P < 0.05$). Blood culture was positive in 3 patients in group A, and 5 in group B. There was no obvious difference in the incidence of infectious complication between the two groups. Two patients died in group A and 1 group B. There was no obvious difference in plasma IL-1 level between A1 and B1 groups at different time points. The plasma IL-6 level in A1 group in 10th and 14th PBD was evidently lower than that in B1 group ($P < 0.05$). The PINI in A1 group on the 10 PBD was remarkably lower than that in B1 group. Early enteral nutrition with synbiotics was helpful in decreasing inflammatory stress response and lowering the plasma endotoxin level. Enteral supplementation of synbiotics might be beneficial to the controlling of burn infection.

The passage of lactic acid bacteria from silage into rumen fluid, in vitro studies; Weinberg ZG et al.; Inoculated silages sometimes improve cattle performance, possibly because of probiotic effects of lactic acid bacteria (LAB) silage inoculants. The cause of improved animal performance following feeding with inoculated silage is unclear. One issue in studying this phenomenon is to find out whether LAB pass from silage into the rumen fluid and survive in it. The purpose of the present study was to determine whether LAB from inoculated and uninoculated silages pass into the rumen fluid in vitro. Wheat and corn silages, uninoculated or inoculated with 1 of 10 commercial silage inoculant LAB, were prepared in glass jars. After ensiling, a 2.5-g silage sample was added to 25 mL of heat-sterilized or strained rumen fluid together with 5 g/L glucose, and incubated for 48 h at 39 degrees C. Analysis of the incubated rumen fluid included pH measurement, enumeration of LAB, and determination of lactic acid and volatile fatty acids (VFA). The pH of the rumen fluid decreased during incubation; both heat-sterilized and strained rumen fluid contained large numbers of LAB. The heat-sterilized rumen fluid contained lactic acid in addition to VFA, whereas the strained rumen fluid contained only VFA. The results indicate that LAB pass from silage samples into the rumen fluid in vitro and survive there. Their interactions with rumen microorganisms should be studied further to understand how some silage inoculant LAB exhibit probiotic effects in dairy cattle.

NMR solution structure of ImB2, a protein conferring immunity to antimicrobial activity of the type IIa bacteriocin, carnobacteriocin B2; Sprules T et al.; Bacteriocins produced by lactic acid bacteria are potent antimicrobial compounds which are active against closely related bacteria. Producer strains are protected against the effects of their cognate bacteriocins by immunity proteins that are located on the same genetic locus and are coexpressed with the gene encoding the bacteriocin. Several structures are available for class IIa bacteriocins; however, to date, no structures are available for the corresponding immunity proteins. We report here the NMR solution structure of the 111-amino acid immunity protein for carnobacteriocin B2 (ImB2). ImB2 folds into a globular domain in aqueous solution which contains an antiparallel four-helix bundle. Extensive packing by hydrophobic side chains in adjacent helices forms the core of the protein. The C-terminus, containing a fifth helix and an extended strand, is held against the four-helix bundle by hydrophobic interactions with helices 3 and 4. Most of the charged and polar residues in the protein face the solvent. Helix 3 is well-defined to residue 55, and a stretch of nascent helix followed by an unstructured loop joins it to helix 4. No interaction is observed between ImB2 and either carnobacteriocin B2 (CbnB2) or its precursor. Protection from the action of CbnB2 is only observed when ImB2 is expressed within the cell. The loop between helices 3 and 4, and a hydrophobic pocket which it partially masks, may be important for interaction with membrane receptors responsible for sensitivity to class IIa bacteriocins.

Antagonistic activity against Helicobacter pylori infection in vitro by a strain of Enterococcus faecium TM39; Tsai CC et al.; Lactic acid bacteria (LAB) strains from infant feces were screened for anti-Helicobacter pylori use. In the beginning, we selected the strains based on their capability to adhere to the human intestinal epithelial cell (Int-407), colonial enterocyte-like Caco-2 cell, human cervical epithelioid carcinoma cell (HeLa), and human gastric carcinoma cell (TSGH 9201). Then, acid and bile salt tolerance of these LAB strains was evaluated. In addition, the ability of these LAB strains to inhibit the growth of H. pylori and to expel H. pylori cells from TSGH 9201 were studied. The spent culture supernatant (SCS) of a selected strain

TM39, i.e., TM39-SCS, significantly inhibited the viability of H. pylori in vitro. It also inhibited the urease activity of H. pylori in vitro. For these antagonistic effects, in addition to pH and lactic acid, some factors in TM39-SCS might play the major role. Treatment of H. pylori with the SCS or cells of strain TM39 significant reduced its binding to TSGH 9201 cells. Although strain TM39 is identified as Enterococcus faecium, it is not vancomycin resistant and is proved to be safe through the invasion study and a 28-day feeding study with Wistar rats.

Expression of the peptidoglycan recognition protein, PGRP, in the lactating mammary gland; Kappeler SR et al.; The peptidoglycan recognition protein, PGRP, known as an intracellular component of neutrophils, has been isolated from camel (Camelus dromedarius) milk by acid precipitation followed by heparin-sepharose affinity chromatography of the supernatant. The mean concentration in milk was about 120 mg/L. It decreased during lactation by 19% and increased in the event of severe mastitis by 45%. The protein bound to lactic acid bacteria and other gram-positive bacteria with an affinity similar to that reported for the human and murine orthologs, although the isoelectric point of the molecule was distinctly higher at pH 9.02. The N-terminus of mature camel PGRP was determined as NH2-ArgGluAspProPro-CO2H. Calculated and measured molecular masses were both 19.1 kDa, excluding the possibility of posttranslational modifcation or binding of cation ligands. The peptide probably builds a homotrimer at high concentration. The corresponding mRNA was isolated from lactating mammary gland tissue, and 5.3 kbp of the corresponding gene was sequenced. Similarities were found to the camel lactoferrin gene with regard to sites of expression and to the region 5' upstream to the gene.

The nucleotide sequence of TTP1, a gene encoding a predicted type II membrane protein; Romero PA et al.; The DNA sequence of a 2967 bp fragment located near the centromere of chromosome II, between the CEN2 and FUR4 genes, was determined. The segment contains a new open reading frame of 1794 bp. The product encoded by the gene, designated TTP1, is a predicted type II membrane protein of 597 amino acid residues with a short cytoplasmic NH2-terminus, a membrane-spanning region and a large COOH-terminal region containing three potential N-glycosylation sites. Gene disruption indicated that TTP1 is not essential for cell growth.

Properties of a recombinant human hemoglobin double mutant: sickle hemoglobin with Leu-88(beta) at the primary aggregation site substituted by Ala; Martin de Llano JJ et al.; A recombinant double mutant of hemoglobin (Hb), E6V/L88A(beta), was constructed to study the strength of the primary hydrophobic interaction in the gelation of sickle Hb, i.e., that between the mutant Val-6(beta) of one tetramer and the hydrophobic region between Phe-85(beta) and Leu-88(beta) on an adjacent tetramer. Thus, a construct encoding the donor Val-6(beta) of the expressed recombinant HbS and a second mutation encoding an Ala in place of Leu-88(beta) was assembled. The doubly mutated beta-globin gene was expressed in yeast together with the normal human alpha-chain, which is on the same plasmid, to produce a soluble Hb tetramer. Characterizations of the Hb double mutant by mass spectrometry, by HPLC, and by peptide mapping of tryptic digests of the mutant beta-chain were consistent with the desired mutations.

The absorption spectra in the visible and the ultraviolet regions were practically superimposable for the recombinant Hb and the natural Hb purified from human red cells. Circular dichroism studies on the overall structure of the recombinant Hb double mutant and the recombinant single mutant, HbS, showed that both were correctly folded. Functional studies on the recombinant double mutant indicated that it was fully cooperative. However, its gelation concentration was significantly higher than that of either recombinant or natural sickle Hb, indicating that the strength of the interaction in this important donor-acceptor region in sickle Hb was considerably reduced even with such a conservative hydrophobic mutation.

Mechanisms of vesicle docking and fusion: insights from the nervous system; Pevsner J et al.; Upon stimulation of nerve cells, synaptic vesicles fuse with the presynaptic plasma membrane to release neurotransmitter. The biochemical pathway responsible for synaptic-vesicle docking and fusion is now being elucidated. Many of the proteins implicated in this process have homologs elsewhere in the cell. The docking and fusion mechanism discussed in this review may account for the specificity of vesicular trafficking throughout both regulated and constitutive secretory pathways. Collaboration of G1 cyclins in the functional inactivation of the retinoblastoma protein; Hatakeyama M et al.; The retinoblastoma gene product (pRB) constrains cell proliferation by preventing cell-cycle progression from the G1 to S phase. Its growth-inhibitory effects appear to be reversed by hyperphosphorylation occurring during G1. This process is thought to involve G1 cyclins and cyclin-dependent kinases (cdks). Here we report that the cell cycle-dependent phosphorylation of mammalian pRB is faithfully reproduced when it is expressed in Saccharomyces cerevisiae. As is the case in mammalian cells, this phosphorylation requires an intact oncoprotein-binding domain and is inhibited by a negative growth factor, in this case a mating pheromone. Expression of pRB in cln (-) mutants indicates that specific combinations of endogenous G1 cyclins, Cln3 and either Cln1 or Cln2 are required for pRB hyperphosphorylation in yeast. Moreover, expression of mammalian G1 cyclins in cln (-) yeast cells indicates that the functions of Cln2 and Cln3 in pRB hyperphosphorylation can be complemented by human cyclin E and cyclin D1, respectively. These observations suggest a functional heterogeneity among G1 cyclin-cdk complexes and indicate a need for the involvement of multiple G1 cyclins in promoting pRB hyperphosphorylation and resulting cell-cycle progression.

Mechanisms of mRNA Degradation in Eukaryotes

Membrane location of apocytochrome c and cytochrome c determined from lipid-protein spin exchange interactions by continuous wave saturation electron spin resonance; Snel MM et al.; Apocytochrome c derived from horse heart cytochrome c was spin-labeled on the cysteine residue at position 14 or 17 in the N-terminal region of the primary sequence, and cytochrome c from yeast was spin-labeled on the single cysteine residue at sequence position 102 in the C-terminal region. The spin-labeled apocytochrome c and cytochrome c were bound to fluid bilayers composed of different negatively charged phospholipids that also contained phospholipid probes that were spin-labeled either in the headgroup or at different positions in the sn-2 acyl chain. The location of the spin-labeled cysteine residues on the lipid-bound proteins was determined relative to the spin-label positions in the different spin-labeled phospholipids by the influence of spin-spin interactions on the microwave saturation properties of the spin-label electron spin resonance spectra. The enhanced spin relaxation observed in the doubly labeled systems arises from Heisenberg spin exchange, which is determined by the accessibility of the spin-label group on the protein to that on the lipid. It is found that the labeled cysteine groups in horse heart apocytochrome c are located closest to the 14-C atom of the lipid acyl chain when the protein is bound to dimyristoyl-or dioleoyl-phosphatidylglycerol, and to that of the 5-C atom when the protein is bound to a dimyristoylphosphatidylglycerol/ dimyristoylphosphatidylcholine (15:85 mol/mol mixture. On binding to dioleoylphosphatidylglycerol, the labeled cysteine residue in yeast cytochrome c is located closest to the phospholipid headgroups but possibly between the polar group region and the 5-C atom of the acyl chains. These data determine the extent to which the different regions of the proteins are able to penetrate negatively charged phospholipid bilayers. Mechanisms of mRNA degradation in eukaryotes; Decker CJ et al.; Recent experiments have identified distinct mechanisms of eukaryotic RNA turnover. In one mechanism, deadenylation triggers decapping, exposing the messenger RNA to 5' to 3' degradation. This pathway may act at different rates on the majority of messenger RNAs. There are also degradation mechanisms, such as endonucleolytic cleavage, limited to messenger RNAs containing specific sequence elements.

Separation of large DNA molecules with high voltage pulsed field gel electrophoresis; Wagner L et al.; We have developed two high voltage (3 kV and 10 kV) high speed pulsed field gel electrophoresis systems for the separation of DNA as large as 460 kbp. These systems enable us to combine the rapid speed of high voltages and the separation power of pulsed field electrophoresis to achieve high resolution and short run durations. We found that large DNA fragments can be separated at voltage gradients much higher than commonly used. Yeast chromosomes as large as 460 kbp can be separated in 4 h at 20 V/cm and 1-50 kbp DNA can be rapidly separated in 30 min at 55 V/cm. This is 25 times faster in mobility for the separation of relatively small DNA fragments (< 50 kbp). We have also found an inverse relationship between the voltage applied and the size separation limit at that particular voltage gradient (55 V/cm limits the separation to 50 kbp while 20 V/cm can separate up to 460 kbp). Depending on the size range, DNA can be separated 8-to 25-fold faster and with better resolution than existing electrophoretic systems.

Mutational analysis of the Sar1 protein, a small GTPase which is essential for vesicular transport from the endoplasmic reticulum; Nakano A et al.; SAR1 encodes a 21-kDa GTPase, which is required for vesicle formation from the endoplasmic reticulum in yeast. Although it belongs to the expanding small GTPase superfamily, there are interesting structural features that are unique to the Sar1 protein. We performed a site-directed mutational study to identify the amino acid residues that are essential for the Sar1p function. Among seven mutants we constructed, four are functionless by themselves, while two confer temperature sensitivity to cells. When the mutant proteins are overproduced in wild-type cells, all of these six show a dominant negative effect on cell growth. The replacement by serine of the only cysteine residue present in Sar1p caused no significant change in the growth phenotype. These findings are not only important for analyzing the mechanism of the Sar1p action in yeast, but will also be very useful for studying the function of Sar1p counterparts in higher eukaryotes.

A structure of oxy-peroxidase as a model for the transient enzyme: peroxide complex; Miller MA et al.; The Fe+3-OOH complex of peroxidases has a very short half life, and its structure cannot be determined by conventional methods. The Fe^{+2}-O_2 complex provides a useful structural model for this intermediate, as it differs by only one electron and one proton from the transient Fe^{+3}-OOH complex. We therefore determined the crystal structure of the Fe+2-O2 complex formed by a yeast cytochrome c peroxidase mutant with Trp 191 replaced by Phe. The refined structure shows that dioxygen can form a hydrogen bond with the conserved distal His residue, but not with the conserved distal Arg residue. When the transient Fe+3-OOH complex is modelled in a similar orientation, the active site of peroxidase appears to be optimized for catalysing proton transfer between the vicinal oxygen atoms of the peroxy-anion.

Regulation of Src family kinases in the developing rat brain: correlation with their regulator kinase, Csk; Inomata M et al.; We have so far suggested that the functions of Src family protein-tyrosine kinases are under the control of their regulator kinase, Csk. To evaluate the role of Csk-mediated regulation in neural tissues, we examined the correlation between the activities of Src family kinases and the expression level of Csk during development of the rat brain. Csk was expressed at high levels in the developing embryonic brain and then rapidly decreased as the brain matured. Consistent with the decrease in the Csk level, the kinase activity of a member of the Src family, Fyn, was greatly enhanced, but that of Src was not correlated inversely with the level of Csk expression. Src exhibited elevated activity in the developing brain, in which a neuronal form of Src (N-Src) is expressed as the dominant form of Src. Although N-Src was readily down-regulated by Csk when coexpressed in yeast, it showed much higher specific activity than c-Src, even in the repressed form. These findings suggest that neural tissues acquire high activities of Src family kinases, which might be important for differentiation and development of the nervous system, through induction of the active form of Src (N-Src) and down-regulation of their suppresser, Csk.

Production and characterization of monoclonal antibodies specific to multi-ubiquitin chains of polyubiquitinated proteins; Fujimuro M et al.; Polyubiquitinated proteins tagged with multi-ubiquitin chains are substrates preferred by the 26 S proteasome (a ubiquitin/ATP-dependent proteolytic complex). Here, we developed a simple method for the efficient preparation of polyubiquitinated proteins which are degraded by the 26 S proteasome in an ATP-dependent manner. Our efficient method enabled us to produce ten monoclonal antibodies that recognized the multi-ubiquitin chains of the polyubiquitinated proteins, but not free ubiquitin or the protein moieties. Eight of the antibodies recognized only the multi-ubiquitin chains of the polyubiquitinated proteins, while the other two antibodies cross-reacted with mono-ubiquitin and methyl-ubiquitin, both of which are linked to proteins via an isopeptide bond, as well as with the multi-ubiquitin chains. Thus these antibodies are novel and useful tools for the identification and quantification of polyubiquitinated proteins in various cells and tissues under physiological and pathological conditions.

Nup145p is required for nuclear export of mRNA and binds homopolymeric RNA in vitro via a novel conserved motif; Fabre E et al.; An essential yeast protein, Nup145p, is identified via its genetic interaction with the nucleoporin Nsp1p. Nup145p contains GLFG repeats and localizes to nuclear pores. Depletion of Nup145p in vivo leads rapidly to nuclear retention of polyadenylated RNAs and more slowly to cytoplasmic accumulation of a nuclear reporter protein. A stretch of 140 amino acids within Nup145p is conserved in two other yeast nucleoporins, Nup116p and Nup100p, and in an uncharacterized C. elegans protein. Genetic experiments in yeast reveal that the three copies of the motif carry out an essential, redundant function. Fragments of Nup145p and Nup116p including this motif bind specifically to homopolymeric RNAs in vitro. Nup145p, Nup116p, and Nup100p thus represent a novel class of nucleoporins involved in nucleocytoplasmic transport.

Characterization of two structurally related Xenopus laevis protein tyrosine phosphatases with homology to lipid-binding proteins; Del Vecchio RL et al.; We have chosen Xenopus laevis as a model system to study how protein tyrosine phosphatases (PTPases) function in growth and development. As an initial step, we have previously isolated in a polymerase chain reaction (PCR)-based protocol cDNA fragments which correspond to sequences within the catalytic domains of PTPases (Yang, Q., and Tonks, N. K. (1993) Adv. Protein Phosphatases 7, 359-372). Two of these PCR products, designated X1 and X10, have now been used to screen a X. laevis ovary cDNA library to obtain complete coding sequences for two distinct PTPases. The X1 cDNA encodes a protein (PTPX1) of 693 amino acids (approximately 79 kDa); the X10 cDNA encodes a protein of 597 amino acids (approximately 69 kDa). Both PTPX1 and PTPX10 lack potential membrane spanning sequences and therefore can be classified as non-transmembrane/cytoplasmic PTPases. While the overall structure of these PTPases are similar, sharing segments of 95% amino acid identity, they differ in that PTPX1 contains a unique 97-amino acid insert between the N-terminal segment and C-terminal catalytic domain. The absence of complete identity between PTPX1 and PTPX10 suggests that these two sequences are the products of separate genes and not the result of alternative splicing. This conclusion is confirmed by PCR analysis of Xenopus genomic DNA. Both PTPases share sequence identities in their N-terminal segments with two lipid-binding proteins, cellular retinaldehyde-binding protein and SEC14p, a phospholipid transferase. In addition, the unique insert sequence of PTPX1 shares identity with PSSA, a protein involved in phosphatidylserine biosynthesis. Sequence comparison suggests that PTPX10 is the Xenopus homolog of the human PTPase Meg-02, while PTPX1 is a structurally related yet distinct PTPase. Intrinsic PTPase activity of PTPX1 and PTPX10 was demonstrated in lysates of Sf9 cells infected with recombinant baculoviruses encoding either enzyme. PTPX1 can be recovered in both soluble and membrane fractions from Xenopus oocytes with the membrane form exhibiting approximately 4-fold higher activity than the soluble form.

Replacement of all alpha-domain lysines with glutamates reduces metallothionein detoxification function; Cody CW et al.; Mammalian metallothioneins (MTs) possess eight highly conserved lysine residues, including three in each of two metal binding domains. We used site-directed mutagenesis to replace these intradomain lysines in Chinese hamster ovary MT2 with glutamic acid and/or glutamine. These mutant MTs were expressed in a metal sensitive yeast host. One mutant which had all three lysines in the alpha-domain replaced by glutamates (K43,51,56E) exhibited a reduced ability, relative to native MT, to protect yeast transformants against otherwise toxic levels of cadmium. This triply substituted mutant also exhibited anomalous migration on a non-denaturing gel relative to wild type MT and other MT lysine mutants, suggesting that the intradomain lysines are important in maintaining the conformational integrity of MT.

Specific interaction of type I receptors of the TGF-beta family with the immunophilin FKBP-12; Wang T et al.; Transforming growth factor-beta (TGF-beta) family members bind to receptors that consist of heteromeric serine-threonine kinase subunits (type I and type II).

In a yeast genetic screen, the immunophilin FKBP-12, a target of the macrolides FK506 and rapamycin, interacted with the type I receptor for TGF-beta and with other type I receptors. Deletion, point mutation, and co-immunoprecipitation studies further demonstrated the specificity of the interaction. Excess FK506 competed with type I receptors for binding to FKBP-12, which suggests that these receptors share or overlap the macrolide binding site on FKBP-12, and therefore they may represent its natural ligand. The specific interaction between the type I receptors and FKBP-12 suggests that FKBP-12 may play a role in type I receptor-mediated signaling.

Catalytically important domains of rat carnitine palmitoyltransferase II as determined by site-directed mutagenesis and chemical modification. Evidence for a critical histidine residue; Brown NF et al.; Rat carnitine palmitoyltransferase (CPT) II was expressed in Saccharomyces cerevisiae. Mitochondrial fractions prepared from the cells displayed high CPT activity and reacted with an antibody to the rat protein on immunoblots, whereas no activity or immunoreactive protein was observed in control cells. The recombinant enzyme was largely membrane associated. Treatment of the expressed protein with diethyl pyrocarbonate, a reagent that modifies histidine residues, abolished CPT activity, but this was completely restored by reversal of the modification with hydroxylamine. It is inferred that a histidine residue plays a critical role in CPT function. Expression and analysis of site-directed mutants of CPT II showed that histidine 372, as well as aspartates 376 and 464 (all conserved throughout the carnitine/choline acyltransferase family), are essential for catalytic activity.

The data suggest that the mechanism by which CPT II effects transesterification between palmitoyl-CoA and carnitine possibly involves histidine 372 and one of these aspartate residues, interacting with the carnitine hydroxyl group, in a reaction analogous to that carried out by a histidine/aspartate/serine catalytic triad in certain other enzyme systems. Mutagenic analysis of a region of CPT II that is highly conserved among the carnitine and choline acyltransferases, and which is homologous to the "adenine binding loop" of citrate synthase, implies that it has no similar function in CPT II. Activation of phosphoinositide 3-kinase activity by Cdc42Hs binding to p85; Zheng Y et al.; The Ras-like GTPase Cdc42 is essential for cell polarity and bud site assembly in Saccharomyces cerevisiae by regulating cell cycle-dependent reorganization of cortical cytoskeletal elements. However, its role in mammalian cells is unknown. To identify potential effectors of Cdc42Hs, we incubated lysates from NIH 3T3 fibroblasts or PC12 cells with immobilized glutathione S-transferase (GST)-Cdc42Hs fusion proteins bound to different guanine nucleotides and observed a specific association between the 85-kDa subunit (p85) of phosphatidylinositol 3-kinase (PI 3-kinase) and GTP gamma S (guanosine 5'-3-O-(thio)triphosphate)-bound GST-Cdc42Hs. Recombinant p85 formed a complex with GTP gamma S-bound GST-Cdc42Hs and with a GTPase-defective GTP-bound GST-Cdc42Hs-Q61L mutant, but not with a GTP gamma S-bound, effector domain GST-Cdc42HsT35A mutant. Both the Rho-GAP homology domain of p85 and the Cdc42Hs-GAP competitively inhibited the binding of recombinant p85 to Cdc42Hs. In addition, PI 3-kinase activity immunoprecipitated from cell lysates with anti-p85 antibody was stimulated 2-4-fold by GST-Cdc42-GTP gamma S. Similar interactions were observed between p85 and GST-Rac1-GTP gamma S but not between p85 and GST-RhoA-GTP gamma S. These findings suggest that PI 3-kinase, through the Rho-GAP homology domain of p85, can couple to the effector domain of Cdc42Hs and that p85 may be a target for the GTP-bound forms of Cdc42Hs and Rac1.

Functional size of C-terminal protein carboxyl methyltransferase from kidney basolateral plasma membranes; Boivin D et al.; The functional sizes of the C-terminal isoprenylcysteine protein carboxyl methyltransferase (PCMT) from kidney cortex basolateral plasma membranes and yeast membranes have been estimated by the radiation inactivation and fragmentation method. Attempts to solubilize the methyltransferase with detergents were unsuccessful as they resulted in the irreversible denaturation of its enzymatic activity. The radiation inactivation sizes of the methyltransferases were 98 and 24 kDa for kidney and yeast, respectively. Kinetic experiments showed that irradiation affects the Vmax of the reaction but not the apparent Km for either S-adenosyl-L-methionine and N-acetyl farnesylcysteine. The functional size reported here for the kidney membrane is about 4-times larger than the size predicted for the Saccharomyces cerevisiae C-terminal PCMT deduced from the nucleotide sequence of its gene (28 kDa). These results suggest that mammalian methyltransferase has a functional size different from that of the yeast; tetramerization of monomers is one possible hypothesis for this difference.

DNA-dependent protein kinase (Ku protein-p350 complex) assembles on double-stranded DNA; Suwa A et al.; The Ku protein is an autoantigen that consists of 70-and 80-kDa polypeptides. It associates with double-stranded DNA at free ends. In the present study, we examined the ability of anti-Ku antibodies to immunoprecipitate various structures from extracts of HeLa cells prepared at different salt concentrations. Under physiological conditions, these antibodies identified a complex containing the Ku protein and the 350-kDa component (p350) of DNA-dependent protein kinase (DNA-PK), which appeared to be closely associated on the DNA strand. In reconstitution experiments with cell extracts and biochemically purified components, the Ku protein-p350 complex formed only in the presence of double-stranded DNA. The reconstituted complex was catalytically active. Together with previous studies, these results indicate that the Ku protein interacts with DNA to create a binding site for p350 as the DNA-PK holoenzyme assembles.

DNA-dependent ATPase from HeLa cells is related to human Ku autoantigen; Cao QP et al.; A 150-kDa DNA-dependent ATPase composed of 83/68-kDa subunits was previously reported to cofractionate with a 21S complex of enzymes for DNA synthesis from HeLa cells (Vishwanatha, J. K., & Baril, E. F. (1990) Biochemistry 29, 8753-8759). The DNA-dependent ATPase was purified to electrophoretic homogeneity from a HeLa cell homogenate by a modified procedure that involves subcellular fractionation, poly(ethylene-glycol) precipitation of the combined nuclear extract/cytosol, and chromatography on Q-Sepharose and native and denatured DNA/ celluloses followed by Mono-S fast protein liquid chromatography. The purified enzyme showed equimolar amounts of 83-and 68-kDa polypeptides following polyacrylamide gel electrophoresis under denaturing conditions. Sequence analysis of peptide fragments derived from the separated 83-and 68-kDa polypeptides showed 90-100% homology with the corresponding 80-and 70-kDa subunits of human Ku protein. Immunoblot analysis of the ATPase during the course of its purification and immunoprecipitation with antibodies to the 80- and 70-kDa subunits of human Ku protein confirmed the relationship of the 83-and 68-kDa polypeptides of the human DNA-dependent ATPase to the subunits of human Ku protein. Both the 83-and 68-kDa polypeptides are phosphorylated by a DNA-dependent protein kinase that cofractionates with the ATPase. The DNA-dependent ATPase activity is up regulated by phosphorylation.

Abnormal gene expression and regulation in the liver of jvs mice with systemic carnitine deficiency; Tomomura M et al.; Carnitine-deficient jvs mice expressed reduced levels of a group of genes which are preferentially expressed in the liver, including urea cycle enzyme genes (Biochim. Biophys. Acta 1138, 167-171, 1992). The expression of alpha-fetoprotein and aldolase A was elevated, indicating that the liver of jvs mice is undifferentiated or dedifferentiated (FEBS Lett. 311, 63-66, 1992). Studies of the hormone signal transduction pathway showed that serum cortisol and plasma glucagon levels of jvs mice were 2 and 3 times higher, respectively, than those of normal mice, and that the hormone binding activity of glucocorticoid receptor (GR) in the cytosol of jvs liver was 50% of normal mice, which reflected the amount of receptor protein in the cytosol.

On the other hand, GR protein accumulated in the nuclear fraction in jvs mice. Exogenously administrated dexamethasone induced carbamoyl phosphate synthetase (CPS) and tyrosine aminotransferase (TAT) mRNAs in jvs mice, indicating that CPS and TAT genes in jvs mice are responsive to induction by glucocorticoid and cAMP. Analysis of transacting factors by gel retardation assay revealed that HNF-1, COUP-TF and SP-1 were detected at almost the same level in the hepatic nuclear fraction of jvs mice as in normal littermates, and C/EBP and CREB were a little higher in jvs mice, suggesting that these factors are probably not targets of jvs mutation causing abnormal gene expression in the liver. On the other hand, AP-1 binding activity was much higher in jvs mice from an early age, preceding the abnormal expression of urea cycle enzyme, and carnitine administration normalized AP-1 binding activity. We suggest that elevated AP-1 binding induced by carnitine deficiency is closely connected with the abnormal gene expression in the liver.

The ABC-transporter Ste6 accumulates in the plasma membrane in a ubiquitinated form in endocytosis mutants; Kolling R et al.; We are investigating the transport and turnover of the multispanning membrane protein Ste6. The Ste6 protein is a member of the ABC-transporter family and is required for the secretion of the yeast mating pheromone a-factor. In contrast to the prevailing view that Ste6 is a plasma membrane protein, we found that Ste6 is mainly associated with internal membranes and not with the cell surface. Fractionation and immunofluorescence data are compatible with a Golgi localization of Ste6. Despite its mostly intracellular localization, the Ste6 protein is in contact with the cell surface, as demonstrated by the finding that Ste6 accumulates in the plasma membrane in endocytosis mutants. The Ste6 protein which accumulates in the plasma membrane in endocytosis mutants is ubiquitinated. Ste6 is thus the second protein in yeast besides MAT alpha 2 for which ubiquitination has been demonstrated. Ste6 is a very unstable protein (half-life 13 min) which is stabilized approximately 3-fold in a ubc4 ubc5 mutant, implicating the ubiquitin system in the degradation of Ste6. The strongest stabilizing effect on Ste6 is, however, observed in the vacuolar pep4 mutant (half-life > 2 h), suggesting that most of Ste6 is degraded in the vacuole. Secretory functions are required for efficient degradation of Ste6, indicating that Ste6 enters the secretory pathway and is transported to the vacuole by vesicular carriers.

Substitution of glutamic acids for the conserved lysines in the alpha domain affects metal binding in both the alpha and beta domains of mammalian metallothionein; Pan PK et al.; Lysine residues are highly conserved in mammalian metallothioneins (MTs). Recombinant mutant Chinese hamster MT2 in which all of the lysines (K) in the alpha-domain were substituted by glutamic acids (E) was assayed with, expressed in and purified from a cadmium sensitive strain of yeast Saccharomyces cerevisiae. Circular dichroism analyses of the mutated protein, mutein K43,51,56E, revealed that the overall structure remained unchanged. However, a 1-D 113Cd NMR study detected significant differences in the chemical shifts of the corresponding resonances between wild type protein and the recombinant mutein. Reduction of integrated intensity in the NMR spectra was also observed for resonances from the four-metal cluster (I and V-VII) in the alpha-domain of the mutein. At various temperatures, facile intermolecular exchange of metals in the beta-domain of the mutein was also observed, which was unexpected and was different from wild type. Our results thus demonstrate that replacing all three lysines by glutamic acids in the alpha-domain changed metal-thiolate interactions in both domains of the recombinant mutein. This may explain the reduced ability of the mutein to convey cadmium resistance. We propose that while the lysine residues in the alpha-domain of wild type MT are not critical for maintaining protein structure, they play a role in regulating the microenvironment and stability of both metal-binding clusters, a feature critical to metal detoxification.

Non-random acetylation of histone H4 by a cytoplasmic histone acetyltransferase as determined by novel methodology; Sobel RE et al.; During periods of active DNA replication and chromatin assembly, newly synthesized histone H4 is deposited in a diacetylated form. In Tetrahymena, a specific pair of residues, lysines 4 and 11, has been shown to undergo this modification in vivo (Chicoine, L. G., Schulman, I. G., Richman, R., Cook, R. G., and Allis, C. D. (1986) J. Biol. Chem. 261, 1071-1076). Presumably, this reaction is catalyzed, at least in part, by histone acetyltransferases (HAT) of the B type, cytoplasmic enzymes displaying strong preference for free, non-chromatin-bound H4. To investigate which lysines are preferred acetylation sites in H4 from other organisms, a cytoplasmic HAT B activity was prepared from Drosophila embryos and used to acetylate H4 from several species. When H4 or synthetic, NH2-terminal peptides from Tetrahymena were used as unblocked substrates, direct microsequence analyses showed that {3H}acetate was preferentially incorporated at lysine 11 with little, if any, incorporation at other conserved, acetylatable lysines. Drosophila H4 was chemically deblocked following its acetylation in vitro using conditions that do not deacetylate internal lysines. Direct sequence analysis verified the correct NH2-terminal sequence of Drosophila H4 and demonstrated that {3H}acetate incorporation occurred preferentially on lysine 12, the residue analogous to lysine 11 in Tetrahymena. These data show remarkable preference for lysine 11/12 by the Drosophila HAT B activity in vitro and provide support for the assertion that this activity functions to acetylate new H4, at least in part, for deposition and chromatin assembly in vivo. Since most H4s, like Drosophila, are blocked at their amino termini by an acetylthreonine or acetylserine, our results demonstrate that this deblocking and microsequencing strategy can be used to study acetylation site utilization in H4 and presumably other core histones NH2 terminally blocked with these residues.

The karyogamy gene KAR2 and novel proteins are required for ER-membrane fusion; Latterich M et al.; We have developed assays using cells and isolated membranes to identify factors mediating fusion of the ER-nuclear membrane network in yeast. When cells containing distinctly tagged ER-nuclear envelope membranes are observed during mating, the markers of both parental membranes become colocalized in a process sharing a genetic requirement with karyogamy. Using isolated membranes, we find that fusion between ER compartments requires ATP, but not cytosol, Sec17p (alpha-SNAP), or Sec18p (NSF), the latter two being required at the fusion step in vesicular transport. Proteins tightly associated with the ER membrane are essential for fusion, as is Kar2p (BiP), an ER lumenal hsp70 homolog. BiP may activate an ER-localized fusogen, allowing nuclear fusion and karyogamy in yeast.

Modulation of Human Dendritic Cell

Transcriptional activation of NFI/CTF1 depends on a sequence motif strongly related to the carboxyterminal domain of RNA polymerase II; Wendler W et al.; Initiation of RNA polymerase II-directed transcription is mediated by DNA sequence specific activator proteins interacting with components of the basal transcription machinery. NFI/CTF is a family of such binding proteins which have been shown to stimulate transcription via proline-rich activation domains. In order to identify residues crucial for its activator function, a pool of CTF1 mutants was cloned and fused to the bacterial repressor LexA. Transcriptional activation of these constructs was monitored in a Saccharomyces cerevisiae reporter assay. Our studies reveal the existence of a core domain in CTF1 between residues 463 and 508 essential for transcriptional activation functions. It contains the sequence motif SPTSPSYSP, which is strongly related to the heptapeptide repeat YSPTSPS present in the carboxyterminal domain (CTD) of RNA polymerase II. Removal of the entire CTD related motif, as well as substitution of key amino acids therein, abolish CTF1 mediated transcriptional activation.

Modulation of human dendritic cell phenotype and function by probiotic bacteria; Hart AL et al.; "Probiotic" bacteria are effective in treating some inflammatory bowel diseases. However which bacteria confer benefit and mechanisms of action remain poorly defined. Dendritic cells, which are pivotal in early bacterial recognition, tolerance induction, and shaping of T cell responses, may be central in mediating the effects of these bacteria. To assess effects of different probiotic bacteria on dendritic cell function. Human intestinal lamina propria mononuclear cells, whole blood, or an enriched blood dendritic cell population were cultured with cell wall components of the eight bacterial strains in the probiotic preparation VSL#3 (four lactobacilli, three bifidobacteria, and one streptococcal strains).

Dendritic cells were identified and changes in dendritic cell maturation/costimulatory markers and cytokine production in response to probiotic bacteria were analysed by multicolour flow cytometry, in addition to subsequent effects on T cell polarisation. VSL#3 was a potent inducer of IL-10 by dendritic cells from blood and intestinal tissue, and inhibited generation of Th1 cells. Individual strains within VSL#3 displayed distinct immunomodulatory effects on dendritic cells; the most marked anti-inflammatory effects were produced by bifidobacteria strains which upregulated IL-10 production by dendritic cells, decreased expression of the costimulatory molecule CD80, and decreased interferon-gamma production by T cells. VSL#3 diminished proinflammatory effects of LPS by decreasing LPS induced production of IL-12 while maintaining IL-10 production. Probiotic bacteria differ in their immunomodulatory activity and influence polarisation of immune responses at the earliest stage of antigen presentation by dendritic cells. Characterization of the groEL and groES loci in Bifidobacterium breve UCC 2003: genetic, transcriptional, and phylogenetic analyses; Ventura M et al.; The bacterial heat shock response is characterized by the elevated expression of a number of chaperone complexes, including the GroEL and GroES proteins. The groES and groEL genes are highly conserved among eubacteria and are typically arranged as an operon. Genome analysis of Bifidobacterium breve UCC 2003 revealed that the groES and groEL genes are located in different chromosomal regions. The heat inducibility of the groEL and groES genes of B. breve UCC 2003 was verified by slot blot analysis. Northern blot analyses showed that the cspA gene is cotranscribed with the groEL gene, while the groES gene is transcribed as a monocistronic unit.

The transcription initiation sites of these two mRNAs were determined by primer extension. Sequence and transcriptional analyses of the region flanking the groEL and groES genes of various bifidobacteria revealed similar groEL-cspA and groES gene units, suggesting a novel genetic organization of these chaperones. Phylogenetic analysis of the available bifidobacterial groES and groEL genes suggested that these genes evolved differently. Discrepancies in the phylogenetic positioning of groES-based trees make this gene an unreliable molecular marker. On the other hand, the bifidobacterial groEL gene sequences can be used as an alternative to current methods for tracing Bifidobacterium species, particularly because they allow a high level of discrimination between closely related species of this genus. Microbiocenosis of the bowels in patients with gastroduodenal ulcers}; Clarification of interactions among microorganisms and development of co-culture system for production of useful substances; Department of Materials Science and Technology, Niigata University, Ikarashi 2-8050, Niigata 950-2181, Japan.

Co-culture systems containing two microorganisms for the production of useful substances are described. We developed a novel co-culture system composed of two fermentors and two microfiltration modules. The proposed co-culture system allowed regulation of the dissolved oxygen concentration at a level suitable for an individual microorganism in each fermentor, as well as the successful exchange of culture medium between two fermentors. By co-culture, using a combination of Pichia stipitis and Saccharomyces cerevisiae, ethanol was produced efficiently from a mixture of glucose and xylose. Moreover, the useful probiotic cells were simultaneously produced with a high productivity by our co-culture using a combination of Bifidobacterium and Propionibacterium. Kefiran production by Lactobacillus kefiranofaciens alone under the culture conditions, established by mimicking the presence and activities of yeast cells in kefir grains, was also investigated. The results obtained showed that under the culture conditions established by mimicking the actions of yeast cells on L. kefiranofaciens in kefir grains, the amount of kefiran produced was enhanced, even when the lactic acid bacterium alone was used.

The protective effect of supplementation of probiotics combined with riboflavin on the intestinal barrier of the rats after scald injury}; Wang ZT et al.; To investigate the effect and its possible mechanism of the supplementation of probiotics combined with riboflavin on the intestinal barriers of the rats after scald injury. Seventy Wistar rats were used in the study and were randomly divided into scald control (SC, n = 30), scald and treatment (ST, n = 30) and normal control (NC, n = 10) groups.

The rats in SC and ST groups were subjected to 30% TBSA III degree scald. 1.5 ml of isotonic saline suspension containing 5 x 10(12) CFU/L of Bifidobacteria, 5 x 10(10) CFU/L of Bacillus cereus and 5 mg/L of riboflavin was given to rats by gavage in ST group twice a day. For the rats in SC and NC group equal amount of isotonic saline was fed twice a day. The changes in the incidence of bacterial translocation, the amount of intestinal membranous flora, the synthesis and secretion of SIgA in the ileum, and the repair of injured intestinal mucosa were observed. The incidence of bacterial translocation in ST group was significantly lower than that in SC group (P = 0.000-0.025). The plasma level of endotoxin in ST group was markedly lower than that in SC group on 3 post-scald day (PSD) (P < 0.05). The amount of bifidobacteria in caecal membrane flora increased by about 20 to 40 fold, whereas the amounts of E. coli and fungi significantly decreased (P < 0.01). The membranous injury scoring was 3 to 0 on 5 PSD (P < 0.05), and the SIgA content in intestinal mucus returned to normal value on the 5th PSD (P < 0.01) in the ST group. Supplementation of probiotics together with riboflavin could ameliorate translocation of bacteria and endotoxin in rats with scald injury, implying that the intestinal barrier function was effectively protected.

Observation on intestinal flora in patients of irritable bowel syndrome after treatment of Chinese integrative medicine}; Sun X et al.; To observe the clinical efficacy of the combination of traditional Chinese medicine and western medicine in treating irritable bowel syndrome (IBS) and the result of intestinal flora regulation. Sixty IBS patients, 36 males and 24 females, were divided into two groups, with 30 patients in each group. Herbal formula of TongxieYaofang and clostridium butyricum (Cb) were used in the first group for four weeks, while only Cb was used for four weeks in the second group. We observed the changes of coliform group, enterococcus, lactobacillus, bifidobacterium after treatment. The effective rate of the Tongxie Yaofang and Cb treated group was significantly higher than that of the Cb treated group ($P < 0.05$). The numbers of bifidobacterium and lactobacillus increased, while the numbers of coliform group and enterococcus decreased after the treatment, and the changes of intestinal flora in the integrative medicine treated group were significantly greater than those in the Cb treated group. After treatment with the combination of traditional Chinese medicine and western medicine, the intestinal flora can be regulated to equilibrium state.

Photochemical treatment of platelet concentrates with amotosalen and long-wavelength ultraviolet light inactivates a broad spectrum of pathogenic bacteria; Lin L et al.; Bacterial contamination of platelet (PLT) concentrates can result in transfusion-transmitted sepsis. A photochemical treatment (PCT) process with amotosalen HCl and long-wavelength ultraviolet light (UVA), which cross-links nucleic acids, was developed to inactivate bacteria and other pathogens in PLT concentrates. High titers of pathogenic aerobic and anaerobic Gram-positive bacteria (10 species), aerobic Gram-negative bacteria (7 species), and spirochetes (2 species) were added to single-donor PLT concentrates containing 3.0 x 10(11) to 6.0 x 10(11) PLTs in approximately 300 mL of 35 percent plasma and 65 percent PLT additive solution (InterSol, Baxter Healthcare) or saline.

After PCT with 150 micro mol per L amotosalen and 3 J per cm(2) UVA, residual bacterial levels were detected by sensitive microbiologic methods. The level of inactivation of viable bacteria was expressed as log reduction. Log reduction of Gram-positive bacteria for Staphylococcus epidermidis was > 6.6; for Staphylococcus aureus, 6.6; for Streptococcus pyogenes, > 6.8; for Listeria monocytogenes, > 6.3; for Corynebacterium minutissimum, > 6.3; for Bacillus cereus (vegetative), > 5.5; for Lactobacillus sp., > 6.4; for Bifidobacterium adolescentis, > 6.0; for Propionibacterium acnes, > 6.2; and for Clostridium perfringens, > 6.5. Log reduction of Gram-negative bacteria for Escherichia coli was > 6.4; for Serratia marcescens, > 6.7; for Klebsiella pneumoniae, > 5.6; for Pseudomonas aeruginosa, 4.5; for Salmonella choleraesuis, > 6.2; for Yersinia enterocolitica, > 5.9; and for Enterobacter cloacae, 5.9. Log reduction of spirochetes for Treponema pallidum was 6.8 to 7.0, and for Borrelia burgdorferi, > 6.9. PCT inactivates high levels of a broad spectrum of pathogenic bacteria. The inactivation of bacteria in PLT concentrates offers the potential to prospectively prevent PLT-transfusion-associated bacteremia.

Bifidobacterium longum requires a fructokinase (Frk; ATP:D-fructose 6-phosphotransferase, EC 2.7.1.4) for fructose catabolism; Caescu CI et al.; Although the ability of Bifidobacterium spp. to grow on fructose as a unique carbon source has been demonstrated, the enzyme(s) needed to incorporate fructose into a catabolic pathway has hitherto not been defined. This work demonstrates that intracellular fructose is metabolized via the fructose-6-P phosphoketolase pathway and suggests that a fructokinase (Frk; EC 2.7.1.4) is the enzyme that is necessary and sufficient for the assimilation of fructose into this catabolic route in Bifidobacterium longum. The B. longum A10C fructokinase-encoding gene (frk) was expressed in Escherichia coli from a pET28 vector with an attached N-terminal histidine tag. The expressed enzyme was purified by affinity chromatography on a Co(2+)-based column, and the pH and temperature optima were determined. A biochemical analysis revealed that Frk displays the same affinity for fructose and ATP (Km(fructose) = 0.739 +/-0.18 mM and Km(ATP) = 0.756 +/-0.08 mM), is highly specific for D-fructose, and is inhibited by an excess of ATP (>12 mM). It was also found that frk is inducible by fructose and is subject to glucose-mediated repression. Consequently, this work presents the first characterization at the molecular and biochemical level of a fructokinase from a gram-positive bacterium that is highly specific for D-fructose.

Antagonistic activities of lactobacilli and bifidobacteria against microbial pathogens; Servin AL; The gastrointestinal tract is a complex ecosystem that associates a resident microbiota and cells of various phenotypes lining the epithelial wall expressing complex metabolic activities. The resident microbiota in the digestive tract is a heterogeneous microbial ecosystem containing up to 1 x 10(14) colony-forming units (CFUs) of bacteria. The intestinal microbiota plays an important role in normal gut function and maintaining host health. The host is protected from attack by potentially harmful microbial microorganisms by the physical and chemical barriers created by the gastrointestinal epithelium. The cells lining the gastrointestinal epithelium and the resident microbiota are two partners that properly and/or synergistically function to promote an efficient host system of defence. The gastrointestinal cells that make up the epithelium, provide a physical barrier that protects the host against the unwanted intrusion of microorganisms into the gastrointestinal microbiota, and against the penetration of harmful microorganisms which usurp the cellular molecules and signalling pathways of the host to become pathogenic. One of the basic physiological functions of the resident microbiota is that it functions as a microbial barrier against microbial pathogens. The mechanisms by which the species of the microbiota exert this barrier effect remain largely to be determined. There is increasing evidence that lactobacilli and bifidobacteria, which inhabit the gastrointestinal microbiota, develop antimicrobial activities that participate in the host's gastrointestinal system of defence. The objective of this review is to analyze the in vitro and in vivo experimental and clinical studies in which the antimicrobial activities of selected lactobacilli and bifidobacteria strains have been documented.

Comparative detection of bacterial adhesion to Caco-2 cells with ELISA, radioactivity and plate count methods; Le Blay G et al.; Different methods are used to study bacterial adhesion to intestinal epithelial cells, which is an important step in pathogenic infection as well as in probiotic colonization of the intestinal tract. The aim of this study was to compare the ELISA-based method with more conventional plate count and radiolabeling methods for bacterial adhesion detection. An ELISA-based assay was optimized for the detection of Bifidobacterium longum and Escherichia coli O157:H7, which are low and highly adherent bacteria, respectively. In agreement with previous investigations, a percentage of adhesion below 1% was obtained for B. longum with ELISA.

However, high nonspecific background and low positive signals were measured due to the use of polyclonal antibodies and the low adhesion capacity with this strain. In contrast, the ELISA-based method developed for E. coli adhesion detected a high adhesion percentage (15%). For this bacterium the three methods tested gave similar results for the highest bacterial concentrations (6.8 Log CFU added bacteria/well). However, differences among methods increased with the addition of decreased bacterial concentration due to different detection thresholds (5.9, 5.6 and 2.9 Log CFU adherent bacteria/well for radioactivity, ELISA and plate count methods, respectively). The ELISA-based method was shown to be a good predictor for bacterial adhesion compared to the radiolabeling method when good quality specific antibodies were used. This technique is convenient and allows handling of numerous samples.

A repeated batch process for cultivation of Bifidobacterium longum; Her SL et al.; A repeated batch process was performed to culture Bifidobacterium longum CCRC 14634. An on-line device, oxidation-reduction potential (ORP), was used to monitor cell growth and uptake of nutrients in the culture. The ORP of the culture medium decreased substantially during fermentation until nutrients were depleted. Six cycles of batch fermentation using ORP as a control parameter were successfully carried out. As soon as ORP remained constant or increased, three-quarters of the broth was removed, and the same volume of fresh medium was fed to the fermenter for a new cycle of cultivation. Average cell concentrations of 1.9 x 10(9) and 3.4 x 10(9) cfu ml(-1) for repeated batch fermentation in MRS (Lactobacilli MRS broth) and WY (containing whey hydrolyzates, yeast extract, l-cysteine) medium, respectively, were achieved. Cell mass productivities for batch, fed-batch and repeated batch fermentation using MRS medium were 0.51, 0.41, and 0.64 g l(-1) h(-1), respectively, and those for batch and repeated batch using WY medium were 0.76, 0.99 g l(-1) h(-1), respectively. The results indicate a possible industrial process to culture Bifidobacteria sp.

Lactobacillus paracasei normalizes muscle hypercontractility in a murine model of postinfective gut dysfunction; Verdu EF et al.; The effects of probiotics on gut dysfunction in postinfective irritable bowel syndrome are unknown. We tested whether probiotics influence persistent muscle hypercontractility in mice after recovery from infection with Trichinella spiralis and analysed the underlying mechanisms. Mice were gavaged with Lactobacillus paracasei, Lactobacillus johnsonii, Bifidobacterium longum, or Bifidobacterium lactis in spent culture medium from days 10 to 21 after infection. Additional mice received heat-inactivated Lactobacillus paracasei, Lactobacillus paracasei-free spent culture medium, or heat-inactivated Lactobacillus paracasei-free spent culture medium.

Lactobacilli enumeration, immunohistochemistry, and cytokine detection (enzyme-linked immunosorbent assay) were performed. Mice were also treated with Lactobacillus paracasei or Lactobacillus paracasei-free spent culture medium from days 18 to 28 after infection. Contractility was measured on days 21 and 28 after infection. Lactobacillus paracasei, but not Lactobacillus johnsonii, Bifidobacterium lactis, or Bifidobacterium longum, attenuated muscle hypercontractility. This was associated with a reduction in the Trichinella spiralis-associated T-helper 2 response and a reduction in transforming growth factor-beta1, cyclooxygenase-2, and prostaglandin

E 2 levels in muscle. Attenuation of muscle hypercontractility by Lactobacillus paracasei-free spent culture medium was abolished after heat treatment. Improvement of muscle hypercontractility at day 28 after infection was also observed after the administration of Lactobacillus paracasei or Lactobacillus paracasei-free spent culture medium from day 18 after infection. Probiotics show strain-dependent attenuation of muscle hypercontractility in an animal model of postinfective irritable bowel syndrome. This likely occurs via both a modulation of the immunologic response to infection and a direct effect of Lactobacillus paracasei or a heat-labile metabolite on postinfective muscle hypercontractility. Lactobacillus paracasei may be useful in the treatment of postinfective irritable bowel syndrome.

Selection of acid tolerant bifidobacteria and evidence for a low-pH-inducible acid tolerance response in Bifidobacterium longum; Takahashi N et al.; Acidity is an environmental condition commonly encountered by lactic acid bacteria and bifidobacteria in the gastrointestinal tract and fermented foods. In the present study, 22 strains of Bifidobacterium were screened for acid tolerance in artificial gastric juice (AGJ, pH 3.0) and fermented milk. AGJ tolerance was found to be strain-specific, with a pronounced variation among the strains. Several strains with a high survival rate in AGJ that belonged to Bifid. longum, Bifid. breve and Bifid. adolescentis were selected. Among them, only strain BL1 of Bifid longum was found to possess a high survival rate in fermented milk during refrigerated storage. Strain BL1 exhibited a survival rate of more than 25% in AGJ at pH 3.0 for 2 h and maintained a viable cfu level of more than 10(8) per gram of product in fermented milk (pH 4.6) under refrigerated conditions for 2 weeks. The acid tolerance of strain BL1 was found to depend on the final growth pH (<4.5). Rapid loss of acid tolerance was observed when the cells were shifted from acid to neutral conditions by addition of NaOH. Strain BL1 cells were able to maintain much higher intracellular pH under acid conditions, in comparison with those of AGJ sensitive mutant (BL1-S) or cells that lost acid tolerance following pH shifting from acid to neutral conditions. These results suggested that a cytoplasmic pH homeostasis system may function in the acid tolerance response in this strain.

Effect of red pepper Capsicum annuum var. conoides and garlic Allium sativum on plasma lipid levels and cecal microflora in mice fed beef tallow; Kuda T et al.; Antihyperlipidemia or hypocholesterolaemic and antibacterial activities of red hot pepper and garlic are well known. To determine the effect of the dietary spices ingested to suppress blood lipids on the intestinal condition, we examined plasma lipid levels and cecal microflora in mice that were fed diets containing 19% (w/w) beef tallow and 2% red pepper Capsicum annuum var. conoides 'Takanotume' (RP) or garlic Allium sativum 'White' (GP) for 4-weeks. Plasma triacylglyceride level was suppressed by the spices. RP lowered cecal bacteroidaceae, a predominant bacterial group (from 9.4 to 9.0 log CFU/g), bifidobacteria (from 8.7 to 7.6 log CFU/g), and staphylococci. Although GP increased the cecal weight including their contents, significant differences were not shown in the cecal microflora. These results suggest that RP can affect the intestinal condition and host health through the disturbance of intestinal microflora.

Dietary nucleic acid and intestinal microbiota synergistically promote a shift in the Th1/Th2 balance toward Th1-skewed immunity; Sudo N et al.; Intestinal microbiota are known to play an important role in the establishment of oral tolerance, thereby protecting the organism from food allergies. Dietary intake of nucleic acid (NA) is also reported to have such an anti-allergic effect; however, one unsolved question is whether or not dietary NA would act through a process of toll-like receptor 9 signaling activated by DNA containing a CpG motif, a well-known sequence leading to immunostimulatory activity. In this study, we focused on the question of whether the addition of dietary NA lacking CpG motifs would allow continued modulation of the Th1/Th2 balance. Germ free (GF) and Bifidobacterium-infantis-monoassociated BALB/c mice were maintained on either an NA-free casein diet or on an NA-supplemented casein diet for 4 weeks. Thereafter, both the in vivo anti-casein antibody levels and in vitro splenocyte cytokine secretion pattern were evaluated. Feeding with a casein diet elicited a substantial increase in the serum anti-casein-specific IgG1, IgG2a, and IgE levels of GF mice fed the NA free-diet. The in vitro cytokine production profile showed that enhanced IL-4 production in the GF mice fed the NA free-diet was markedly reduced by the supplementation with dietary NA in both the GF and B.-infantis-monoassociated mice. In addition, IFN-gamma secretion increased in the B.-infantis-reconstituted mice fed the diet containing NA. These results suggest that dietary intake of NA devoid of CpG motifs may prevent the development of allergies via acceleration of Th1-dominant immunity.

Cloning and characterization of the bile salt hydrolase genes (bsh) from Bifidobacterium bifidum strains; Kim GB et al.; Biochemical characterization of the purified bile salt hydrolase (BSH) from

Bifidobacterium bifidum ATCC 11863 revealed some distinct characteristics not observed in other species of Bifidobacterium. The bsh gene was cloned from B. bifidum, and the DNA flanking the bsh gene was sequenced. Comparison of the deduced amino acid sequence of the cloned gene with previously known sequences revealed high homology with BSH enzymes from several microorganisms and penicillin V amidase (PVA) of Bacillus sphaericus. The proposed active sites of PVA were highly conserved, including that of the Cys-1 residue. The importance of the SH group in the N-terminal cysteine was confirmed by substitution of Cys with chemically and structurally similar residues, Ser or Thr, both of which resulted in an inactive enzyme. The transcriptional start point of the bsh gene has been determined by primer extension analysis. Unlike Bifidobacterium longum bsh, B. bifidum bsh was transcribed as a monocistronic unit, which was confirmed by Northern blot analysis. PCR amplification with the type-specific primer set revealed the high level of sequence homology in their bsh genes within the species of B. bifidum.

Characterization and heterologous expression of the oxalyl coenzyme A decarboxylase gene from Bifidobacterium lactis; Federici F et al.; Oxalyl coenzyme A (CoA) decarboxylase (Oxc) is a key enzyme in the catabolism of the highly toxic compound oxalate, catalyzing the decarboxylation of oxalyl-CoA to formyl-CoA. The gene encoding a novel oxalyl-CoA decarboxylase from Bifidobacterium lactis DSM 10140 (oxc) was identified and characterized. This strain, isolated from yogurt, showed the highest oxalate-degrading activity in a preliminary screening with 12 strains belonging to Bifidobacterium, an anaerobic intestinal bacterial group largely used in probiotic products. The oxc gene was isolated by probing a B. lactis genomic library with a probe obtained by amplification of the oxalyl-CoA decarboxylase gene from Oxalobacter formigenes, an anaerobic bacterium of the human intestinal microflora.

The oxc DNA sequence analysis revealed an open reading frame of 1,773 bp encoding a deduced 590-amino-acid protein with a molecular mass of about 63 kDa. Analysis of amino acid sequence showed a significant homology (47%) with oxalyl-CoA decarboxylase of O. formigenes and a typical thiamine pyrophosphate-binding site that has been reported for several decarboxylase enzymes. Primer extension experiments with oxc performed by using RNA isolated from B. lactis identified the transcriptional start site 28 bp upstream of the ATG start codon, immediately adjacent to a presumed promoter region. The protein overexpressed in Escherichia coli cross-reacted with an anti-O. formigenes oxalyl-CoA decarboxylase antibody. Enzymatic activity, when evaluated by capillary electrophoresis analysis, demonstrated that the consumption substrate oxalyl-CoA was regulated by a product inhibition of the enzyme. These findings suggest a potential role for Bifidobacterium in the intestinal degradation of oxalate. Efficacy of Leuconostoc mesenteroides (ATCC 13146) isomaltooligosaccharides as a poultry prebiotic; Chung CH et al.; The complex dietary carbohydrates, called prebiotics, have been used to control Salmonella and improve intestinal bacterial balance in broilers. Leuconostoc mesenteroides (ATCC 13146) isomaltooligosaccharides (IMO) stimulate growth of Bifidobacterium and Lactobacillus and are not used by Salmonella or Escherichia coli. We tested the efficacy of these IMO as a prebiotic. IMO, compared with fructooligosaccharides (FOS) as sole carbon source, promoted growth of chicken cecal isolates and Bifidobacterium. Cecal isolates and Salmonella typhimurium grown in mixed culture on IMO reduced the Salmonella population. Cecal isolates grown on IMO showed higher viable counts and faster growth than Salmonella, indicating a potential value for these oligomers for poultry intestinal microflora modification.

Expression, purification, crystallization and preliminary X-ray diffraction analysis of conjugated bile salt hydrolase from Bifidobacterium longum; Kumar RS et al.; Conjugated bile salt hydrolase (BSH) catalyses the hydrolysis of the amide bond that conjugates bile acids to glycine and to taurine. The BSH enzyme from Bifidobacterium longum was overexpressed in Escherichia coli BL21(DE3), purified and crystallized. Crystallization conditions were screened using the hanging-drop vapour-diffusion method. Crystal growth, with two distinct morphologies, was optimal in experiments carried out at 303 K. The crystals belong to the hexagonal system, space group P622 with unit-cell parameters a = b = 124.86, c = 219.03 A, and the trigonal space group P321, with unit-cell parameters a = b = 125.24, c = 117.03 A. The crystals diffracted X-rays to 2.5 A spacing. Structure determination using the multiple isomorphous replacement method is in progress.

The genes for ribosomal RNA have changed little over millions of years as organisms evolved. The slight changes that have occurred provide clues as to how closely or distantly various organisms are related. Because the 16S rRNA gene is very short, just 1,542 nucleotide bases, it can be quickly and cheaply copied and sequenced. So when a scientist has a test tube full of pond water or dirt from an arid mountainside, she must first pull out the rRNA that's mixed up with all the other RNA, DNA and other stuff in that tube. To do this, she cleans and purifies the sample first, getting rid of unwanted debris.

She then uses one or several techniques designed to break open cells like a kid cracking open a piggybank. Now she has to find the 16S rRNA genes in and amongst all the other genes. Although 16S rRNA genes from different microbes will have a few different nucleotides scattered throughout the sequences, those nucleotides at the very beginning or end of the gene are the same from organism to organism. The scientist uses several copies of another bit of RNA called a primer. A primer is like a mirror image of a short bit of RNA or single strand of DNA; that is, its sequence of nucleotides is the direct complement to the sequence of nucleotides in a known part of the target RNA or DNA.

In this instance, the primer would be the mirror image of the beginning or end of the 16S rRNA sequence. Because complementary nucleotides pair up like the two halves of Velcro, the primer enables the scientist to pick out the 16S rRNA in the sample. The scientist then uses PCR to make millions of copies of these genes. She then has enough 16S rRNA to compare the sequences of the genes from her sample to libraries of stored 16S rRNA genes from numerous known bacteria. If some of her gene sequences match up perfectly, she knows that these are microbes that have been previously identified. But if others among her sampled sequences show differences, she knows she has found previously unknown microbes. Bacteria can be found virtually everywhere. They are in the air, the soil, and water, and in and on plants and animals, including us. A single teaspoon of topsoil contains about a billion bacterial cells (and about 120,000 fungal cells and some 25,000 algal cells). The human mouth is home to more than 500 species of bacteria. Some bacteria (along with archaea) thrive in the most forbidding, uninviting places on Earth, from nearly-boiling hot springs to super-chilled Antarctic lakes buried under sheets of ice. Microbes that dwell in these extreme habitats are aptly called extremophiles. Like dinosaurs, bacteria left behind fossils. The big difference is that it takes a microscope to see them. And they are older. Bacteria and their microbial cousins the archaea were the earliest forms of life on Earth. And may have played a role in shaping our planet into one that could support the larger life forms we know today by developing photosynthesis.

Cyanobacteria fossils date back more than 3 billion years. These photosynthetic bacteria paved the way for today's algae and plants. Cyanobacteria grow in the water, where they produce much of the oxygen that we breathe. Once considered a form of algae, they are also known as blue-green algae. The human body consists of millions of different cells. A bacterium consists of a single cell. A bacterium's genetic information is contained in a single DNA molecule suspended in a jelly-like substance called cytoplasm. In most cases, this and other cell parts are surrounded by a flexible membrane that is itself surrounded by a tough, rigid cell wall. A few species, such as the mycoplasmas, don't have cell walls. Even though bacteria have only one cell each, they come in a wide range of shapes, sizes, and colors.

Does a bacterium's cell wall, shape, way of moving, and environment really matter?

Yes! The more we know about bacteria, the more we are able to figure out how to make microbes work for us or stop dangerous ones from causing serious harm. And, for those of us who like to ponder more philosophical questions like the origins of the Earth, there may be some clues there as well. Whether a bacteria has a thin or a thick cell wall determines what antibiotic will work against it. If you've ever been sick and waited for the results of a culture and sensitivity test, you may have heard the terms "Gram-positive" or "Gram-negative."

Bacteria with thick cell walls retain dye from a cell-staining method developed by Christian Gram; bacteria with thin walls do not. Knowing the difference can and does save lives, time and money, and ensures that you or your loved one is getting the best and most effective treatment. Some bacteria look like little balls (Micrococcus) while others appear like tangled strings or corkscrews (Leptospira) under a microscope. Others look like medicine capsules (Salmonella) or segmented ribbons (Cyanobacteria) or sticks. Still others look like fat commas (Vibrios). Some bacteria are stalked (Caulobacter) while others have buds (Rhodomicrobium). Some have sheaths (Sphaerotilus) while others don't. Bacteria like mycoplasmas that lack a hard cell wall don't have any particular shape at all.

Just like in animals, where size ranges from the giant blue whale to the tiny gnat, bacteria vary from 1 millimeter in diameter at the largest end of the scale to 20 nanometers in length at the smallest.

The largest bacteria found so far can actually be seen without the use of a microscope (Thiomargarita namibiensis and Epulopiscium fischelsoni). The smallest known bacteria are so tiny that they were once thought to be viruses (Mycoplasmas). Some bacteria have hair- or whip-like appendages called flagella used to 'swim' around. Others produce thick coats of slime and 'glide' about. Some stick out thin, rigid spikes called fimbriae to help hold them to surfaces. Some contain little particles of minerals that orient with the planet's magnetic fields to help the bacteria figure out whether they're swimming up or down.

Fungi are eukaryotic organisms. This means that their DNA-containing chromosomes are enclosed within a nucleus inside their cells. (The chromosomes of bacteria and archaea are not walled off inside nuclei, making them prokaryotic organisms.) Many decades ago, scientists thought that fungi were primitive kinds of plants. New studies looking at the DNA of fungi have confirmed that these organisms are not plants.

Unlike plants, fungi do not make their own food energy via photosynthesis, but dine on organic matter like rotting leaves, wood, and other debris, or upon the tissues of living plants and animals. Fungi, along with bacteria, are the planet's major composters and recyclers.

Alternative Pasteurization Standards and Raw Milk

In addition to the standard HTST and UHT pasteurization standards, there are other lesser-known pasteurization techniques. The first technique, called "batch pasteurization", involves heating large batches of milk to a lower temperature, typically 155 degrees fahrenheit (or 68 C). The other technique is called higher-heat/shorter time (HHST), and it lies somewhere between HTST and UHT in terms of time and temperature.

The batch pasteurization step, which is cheap at a large scale, is often performed prior to standard pasteurization. Batch pasteurized milk is often called "raw milk" or, confusingly, "unpasteurized milk". It cannot be called "pasteurized", even though a significant number of pathogens are destroyed during the process.

In recent years, there has been some consumer interest in raw milk products, due to perceived health benefits. Advocates of raw milk maintain, correctly, that vitamins and nutrients survive much better in milk that has not been pasteurized. They also maintain that organic raw milk (most retail raw milk is also organic) is less likely to contain harmful pathogens due to better husbandry in organic dairy herds. This may be true, but it has not been proven.

However, doctors (and even most raw milk advocates) acknowledge that certain people (e.g. pregnant or breast-feeding mothers, those undergoing immunosuppression treatment for cancer, organ transplant or autoimmune diseases, and those who are immunocompromised due to diseases like AIDS) should not risk consumption of raw milk.

In fact, some doctors suggest that babies and breast-feeding mothers avoid all but UHT pasteurized dairy products.

In Africa, it is common to boil milk whenever it is harvested. This intense heating greatly changes the flavour of milk, which the people in Africa are accustomed to. Although fungi may seem like a nuisance when they dine in your fruit bowl or refrigerator, their ability to degrade some of the toughest organic materials, including tree wood and insect exoskeletons, means that our planet is not cluttered with a mass of debris. Fungi dine at home, eating whatever they're growing on. Fungi secrete digestive enzymes in order to break down complex food sources, such as animal corpses and tree stumps, into smaller components they can absorb. Algae are plant-like microorganisms that preceded plants in developing photosynthesis, the ability to turn sunlight into energy. Algae cells contain light-absorbing chloroplasts and produce oxygen through photosynthesis.

Although plants generally get the credit for producing the oxygen we breathe, some 75% or more of the oxygen in the planet's atmosphere is actually produced by photosynthetic algae and cyanobacteria. Algae also play an important role as the foundation for the aquatic food chain. All higher aquatic life forms depend either directly or indirectly on microscopic gardens of algae. Most unicellular algae live in water, some dwell in moist soil, and others join with fungi to form lichens.

Green Algae

The most clearly plant-like algae, this species gets its namesake hue from high levels of chlorophyll. Their cell walls are made up of cellulose, the same material that makes up the cell walls in larger, multicellular plants. Like plants, they store the food they make through photosynthesis as starches. Growing in large masses, these algae can form visible layers of slick, green scum on the surfaces and sides of ponds, puddles or damp soil. Fossil records suggest that the first green algae originated 500 to 600 million years ago. Early algae probably gave rise to multicellular plants.

Dinoflagellates have long whip-like structures called flagella that let them turn, maneuver and spin about through the water. About 90% of these algae dwell in the ocean. Some species glow in the dark in a process called bioluminescence. These species contain a compound called luciferin (the same compound found in fireflies). The glow increases markedly if the algae cells are agitated, as when a ship churns through the water. About half the species of dinoflagellates are photosynthetic; the other half are predators that attack bacteria, algae, and even fish. Dinoflagellate neurotoxins can concentrate in the bodies of shellfish and fish that eat the algal cells, in turn causing people who eat these sea foods to come down with illnesses such as paralytic shellfish poisoning and ciguatera (a combination of gastrointestinal, neurological, and cardiovascular disorders.)

So-called "red tides" occur when enormous blooms of trillions of dinoflagellates are triggered by an upwelling of nutrients from the water's depths during warmer seasons. The population of dinoflagellates can jump to more than 20 million cells per liter of sea water along some coasts during these blooms, turning the water a reddish hue. The name protozoa means "first animals." As the principal hunters and grazers of the microbial world, protozoa play a key role in maintaining the balance of bacterial, algal, and other microbial life. They also are themselves an important food source for larger creatures and the basis of many food chains.

Protozoa have been found in almost every kind of soil environment from peat bogs to arid desert sands. They teem in the deep sea as well as near the surface of waters, and can be found even in frigid Arctic and Antarctic waters. Some species of protozoa are part of the normal microbial flora of animals, and live in the guts of insects and mammals, helping to break down complex food particles into simpler molecules. A very small number of species cause disease in people, including Plasmodium vivax, which causes malaria.

Bacterial infection is the most common cause of food poisoning. In the United Kingdom during 2000 the individual bacteria involved were as follows: Campylobacter jejuni 77.3%, Salmonella 20.9%, Escherichia coli O157:H7 1.4%, and all others less than 0.1%. Symptoms for bacterial infections are delayed because the bacteria need time to multiply. They are usually not seen until 12-36 hours after eating contaminated food.

Common bacterial foodborne pathogens are:

Aeromonas hydrophila, Aeromonas caviae, Aeromonas sobria Bacillus cereus Brucella spp. Campylobacter jejuni which causes Guillain-Barré syndrome Corynebacterium ulcerans Coxiella burnetii or Q fever Crohn's disease Escherichia coli O157:H7 enterohemorrhagic (EHEC) which causes hemolytic-uremic syndrome Escherichia coli-enteroinvasive (EIEC) Escherichia coli-enteropathogenic (EPEC) Escherichia coli-enterotoxigenic (ETEC) Listeria monocytogenes

Plesiomonas shigelloides Salmonella spp. Shigella spp. Streptococcus Vibrio cholerae, including O1 and non-O1 Vibrio parahaemolyticus Vibrio vulnificus Yersinia enterocolitica and Yersinia pseudotuberculosis. The four main subgroups of protozoa are the ciliates, the flagellates, the sarcodina, and the apicomplexans.

A few ciliates can grow up to 2 millimeters in length, big enough to be seen without a microscope.

Flagellates Similarly complex single-celled organisms, flagellates have whip-like appendages called flagella sticking out of their cells. The flagella are used for locomotion and to direct food particles or cells into the organism's mouth-like opening. Flagellates dine on bacteria, algae, and other protozoa. Several well-known flagellates cause parasitic diseases, such as trypanosomes that cause sleeping sickness, and Giardia lamblia, a parasite found in mountain streams and rivers that causes severe gastrointestinal distress.

Heliozoa, radiozoa, and forams tend to be passive grazers and predators, relying on suitable food swimming or drifting past to come into contact with their pseudopods. Several species in the sarcodina group, including some species of amoebas, cover themselves with protective shell-like coverings called tests. These tests are stippled with many small and large openings through which water can flow in and out and through which the pseudopods protrude. The tests of radiozoa are made up of silica (the same substance in diatom cell walls) and can form very intricate, lacy designs that may be studded with long spines that increase buoyancy and ward off predators. The tests of forams are made up of sand grains or organic compounds. These can become quite large, the biggest reaching a little over 2 inches in diameter, making them some of the largest single-celled organisms known. When forams die, their tests sink and accumulate in large batches; the Great Pyramids of Egypt are built from sandstone composed largely of fossilized giant Nummulites, an ancient kind of foram. The famous White Cliffs of Dover are limestone cliffs formed from the skeletal remains of forams.

Apicomplexans

These protozoa are obligatory intracellular parasites: they must spend at least part if not all of their life cycle in a host animal. Apicomplexans are characterized by the presence of special organelles (tiny organ-like structures) located at the tips (apices) of the cells. These organelles contain enzymes that punch through, slice open and otherwise penetrate host tissues. The best known apicomplexan is

Plasmodium, the agent that causes malaria. Plasmodium spends part of its life cycle in mosquitoes and the other part in human hosts where it ultimately infects and ruptures blood cells in large numbers.

Another familiar apicomplexan is Cryptosporidium parvum. This water-borne parasite forms extremely durable cyst-like structures that enable it to survive UV radiation and sometimes chlorine in swimming pools and treated water. An outbreak of Cryptosporidium in Milwaukee's drinking water supply in 1993 killed 50 people and sickened more than 400,000. Probably the best known and most deadly case of contamination in the U.S. in recent years happened in Milwaukee in 1993, when the municipal water supply was contaminated by Cryptosporidium, an intestinal protozoan. At least 50 people died, and some 400,000 people became ill, 4,000 badly enough to be hospitalized. In 1973, a Dallas resident went out to the backyard only to stumble upon a reddish, jelly-like mass pulsating in the grass. News reports on the discovery claimed that a "new life form" had been found, and many people couldn't help recalling the cult classic sci-fi thriller The Blob.

Scientists called to the scene, however, put any fears of menacing goo or alien creatures to rest by identifying the mass as an unusually large (46 centimeters or more than 14 inches in diameter) plasmodial slime mold. Slime molds were once considered fungi, but unlike fungi, they can move, and their cell membranes are made of different stuff. Slime molds are made up of individual cells that form an aggregate mass. In their visible, aggregate states, they look like blobs, gooey or foamy masses, spilled jelly, or even dog vomit. They may be bright orange, red, yellow, brown, black, blue, or white.

These large masses act like giant amoebas, creeping slowly along and engulfing food particles along the way. If a slime mold aggregate is diced up, the pieces will pull themselves back together. The blobs can navigate and avoid obstacles and if a food source is placed nearby, they seem to sense it and head unerringly for it. There are two kinds of slime molds. Plasmodial slime molds (the most common kind) share one big cell wall that surrounds thousands or millions of nuclei. Proteins called microfilaments act like tiny muscles that enable the mass to crawl at rates of about 1/25th of an inch per hour.

As long as there is enough food and moisture, the mass thrives. But when food and water are scarce, the mass separates into smaller blobs. The Plasmodium forms stalks topped by sphere-like fruiting bodies that contain spores that are carried by the rain or wind to new locations. Cellular slime molds also produce spores, but these germinate into amoeba-like cells. The cells happily go their individual ways, as long as food and water are available. When nutrients and moisture are scarce, individual cells send out a chemical beacon to attract other cells of the same species. The cells join up to form a mass that looks and acts like a slug to take them to a more favourable location.

Cells in cellular slime molds retain their individual cell walls when they form a mass, so the visible slug is actually a collection of hundreds of thousands of individual cells joined together. Slime molds eat decaying vegetation, bacteria, fungi, and even other slime molds. They are most commonly found in forests. Milk pasteurization standards have been subject to increasing scrutiny in recent years, due to the discovery of pathogens that are both widespread and heat resistant (able to survive pasteurization in significant numbers). Researchers have developed more sensitive diagnostics, such as real-time PCR and improved culture methods, that have enabled them to identify pathogens in pasteurized milk.

Note: The following paragraphs in this section discuss controversial, ongoing research.

One bacterium in particular, the organism Mycobacterium avium subspecies paratuberculosis (MAP), which causes Johne's disease in cattle and is suspected of causing at least some Crohn's disease in humans, has been found to survive pasteurization in retail milk in the U.S., the U.K., Greece, and the Czech Republic. The food safety authorities in the U.K. have decide to re-evaluate pasteurization standards in light of the MAP results and other evidence of harmful, pasteurization-resistant pathogens.

The USDA (which is responsible for setting pasteurization stardards in the U.S.) has not re-evaluated their position on pasteurization adequacy. They do not dispute the studies, which are at this point accepted by the scientific community, but maintain that the presence of MAP in retail pasteurized milk must be due to post-pasteurization contamination. However, some researchers within the FDA, which is responsible for food safety in the U.S., have begun pushing for a re-evaluation of these results. There is a small but growing body of criticism directed at these agencies by Crohn's disease sufferers, scientists, and doctors. Some have suggested that the U.S. dairy industry has been successful in suppressing the agencies' response to a potential health crisis, for fear of consumer panic which would lead to a decrease in milk consumption. It is worth noting that while

MAP has not been definitely proven to be harmful in humans, all other mycobacteria are pathogenic, and it has been definitively shown to cause disease in cattle and other ruminants. The term cold pasteurization is used sometimes for the use of radioactivity or other means (e.g. chemical) to kill bacteria in food.

Water molds are always found in wet environments, especially in fresh water sources and near the upper layers of moist soil. Officially named Oomycota, they are also known as downy mildews and white rusts. Water molds were long considered fungi because they produce fungi-like filamentous hyphae and feed on decaying tissue like rotting logs and mulch. The Oomycota species Phytophthora infestans caused the Great Potato Famine that killed nearly a million people in Ireland in 1846–1847. The water mold virtually wiped out the country's potato crops, which were an essential staple in the Irish diet.

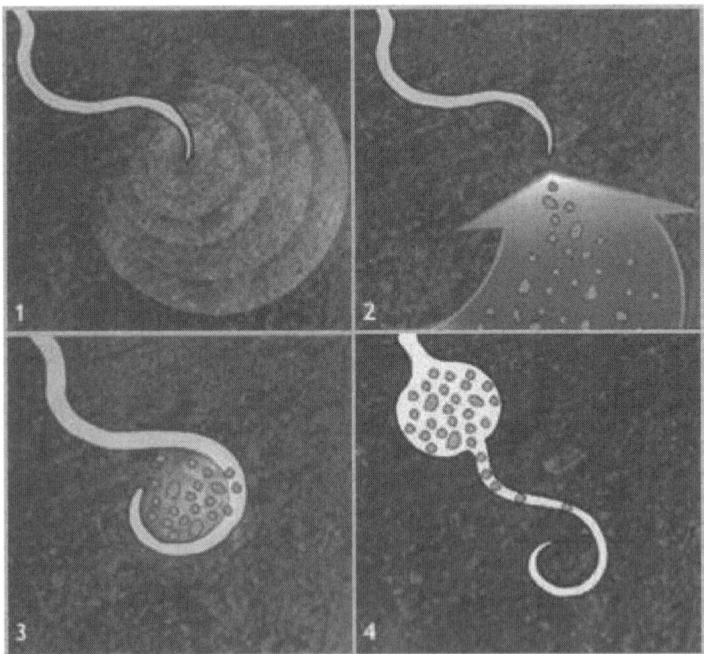

Figure: *The pea root sends a chemical message that attracts Rhizobia, then surrounds the bacteria, whcih set up housekeeping inside the root's cells.*

Microbes break down food molecules our body's enzymes and acids can't dissolve, helping us squeeze all the nutrients out of our food. Some make valuable vitamins that our body needs. Many microbial species have proved to be consummate evolutionary wheelers and dealers, arranging collaborations, mergers, and acquisitions that usually serve both partners well. If you could peer deep into one of the many cells in your body, you'd see little blobs, squiggles, and coils. These are the cell's organelles, structures that perform specialized functions in cells the same way that the lungs, heart, and other organs do in a body.

Mitochondria are the energy factories found in each cell of fungi, protozoa, insects, and animals. Once nutrients are absorbed or digested, they move in the form of minuscule molecules into the mitochondria, which convert the molecules into chemical energy to power the cell.

Chloroplasts undertake a similar function in the cells of plants, algae, and some protozoa. They capture sunlight and, through a series of chemical reactions called photosynthesis, use the light to make energy. These organelles are absolutely essential to the existence of all higher life forms on Earth. If all of the mitochondria in our bodies were to suddenly shut down, we would die. The same is true for plants were they to lose their chloroplasts.

So where did they come from?

As microscopes improved over the years, scientists began noticing striking similarities in the appearances of mitochondria, chloroplasts, and bacterial cells. They discovered that these two organelles contain their own DNA, or gene set, organized very much like the DNA in bacterial cells. Mitochondria and chloroplasts also reproduce independently from the cells in which they reside, in a manner very like bacterial fission. Many microbiologists think it is likely that mitochondria and chloroplasts were once free-living prokaryotes (cells that lack a nucleus and organelles) that somehow took up residence in larger cells. An Invading bacterium may have infected a cell, then become a permanent resident as it adapted to become less virulent, or the target cell became less susceptible. Or both.

Chloroplasts likely first entered a host cell as food before establishing a successful merger with the cell so they were not digested. Ages ago, as land plants were evolving, they ran into a few impediments. Soil can sometimes prove a nutrient-poor and inhospitable environment. In order to grow and thrive, plants need nitrogen to make proteins, but they lack the chemistry set to convert free nitrogen in the air into a form their cells can use. To overcome these obstacles, early plants struck deals with co-evolving bacteria and fungi.

Some early bacteria developed the chemical tools to harvest free nitrogen from the air and convert it into forms such as ammonia and nitrate through a process called nitrogen fixation. The catch is that they need sufficient amounts of energy in the form of carbohydrates to power these conversions, and the supply of carbohydrates in the soil can be limited. On the other hand, plants produce copious amounts of carbohydrates as a product of photosynthesis. And ammonia and nitrate are perfect protein-building nitrogen forms for plants.

The bacteria moved into the plants' roots, forming bumps on the roots called nodules that supply the fixed nitrogen plants need. In exchange, the plants supply the bacteria with the carbohydrates they require. Because Rhizobia can still dwell independently in the soil, plants are more dependent upon them than the microbes are on plants. The prevention is mainly the role of the state, through the definition of strict rules of hygiene and a public service of veterinary survey of the food chain, from farming to the transformation industry and the delivery (shops and restaurants).

This regulation includes: traceability: in a final product, it must be possible to know the origin of the ingredients (originating farm, identification of the harvesting or of the animal) and where and when it was processed; the origin of the illness can thus be tracked and solved (and possibly penalized), and the final products can be removed from the sale if a problem is detected; respect of hygiene procedures like HACCP and the "cold chain"; power of control and of law enforcement of the veterinarians. At home, the prevention mainly consists of: the respect of the food storage and food preservation methods (especially refrigeration), and checking the expiration date; washing the hands before preparing the meal and before eating; washing the fresh vegetables with clear water, especially when not cooked (e.g. fruits, salads); washing the dishes after use; keeping the kitchen clean. Bacteria need warmth, moisture, food and time to grow. The presence, or absence, of oxygen, salt, sugar and acidity are also important factors for growth. In the right conditions, one bacterium can multiply using binary fission to become four million in eight hours. Since bacteria can be neither smelled nor seen, the best way to ensure that food is safe is to follow principles of good food hygiene. This includes not allowing raw or partially cooked food to touch dishes, utensils, hands or work surfaces previously used to handle even properly cooked or ready to eat food.

High salt, high sugar or high acid levels keep bacteria from growing, which is why salted meats, jam, and pickled vegetables are traditional preserved foods. The most frequent causes of bacterial foodborne illness are cross-contamination and inadequate temperature control. Therefore control of these two matters is especially important.

Thoroughly cooking food until it is piping hot, i.e. above 70°C (158°F) will quickly kill virtually all bacteria, parasites or viruses, except for Clostridium botulinum and Clostridium perfringens, which produces a heat-resistant spore that survives temperatures up to 100°C (212°F). Once cooked, hot foods should be kept hot-above 63°C (145°F) stops microbial growth.

Cold foods should be kept cold, below 5°C (41°F). However, Listeria monocytogenes and Yersinia enterocolitica can both grow at refrigerator temperatures. Animals need nitrogen for protein building, too. We humans get our nitrogen by eating plants (or by eating animals that eat plants). Another partnership teams plants with soil-dwelling fungi called mycorrhizae. Virtually all plants from flowers to towering trees like Sequoias have partner mycorrhizae.

Some species of mycorrhizae cover the surface of plants' root hairs; others settle down inside the plant roots. The fungi act as extensions of the plants' roots, vastly increasing the surface space of their nutrient-absorbing network. Mycorrhizae increase the plants' uptake of water and essential nutrients, particularly phosphorous, which doesn't spread readily in soil. In exchange, the plants provide the fungi energy in the form of carbohydrates. This partnership enables both plants and fungi to survive in nutrient-poor places where they otherwise might die. Major development projects are taking place in oceans across the globe all the time, enterprises that will provide shelter and food for a vast number of fish, mussels, urchins, and other marine life. While credit is regularly and duly given to the visible construction crew — coral polyps — recognition is also due the polyps' invisible, but very active algal partners, the zooxanthelle.

These algae (a type called dinoflagellates) live inside the body tissues of coral polyps. Coral polyps take care of some of their nutritional needs on their own by catching tiny protists and organic matter that drift past their tentacles. Their partnership with the chlorophyll-containing algae enable them to also get food from sunlight as if they were plants. The zooxanthelle do the actual work of converting the sunlight into energy via photosynthesis. The by-products they generate (organic carbons such as glycerol and sugars) are excellent nutrients for their polyp hosts.

The zooxanthellae supply much of the polyps' energy needs. In turn, the polyps provide the algae a protected, stable environment and nutrients they need for growth, such as nitrates and phosphates. Zooxanthellae also increase the rate of coral calcification, the growth of the hard shells that form the actual reef structures.

Ants lack the necessary enzymes to digest leaves and stems themselves, so they cut them up and feed them to the fungi, which break down cellulose (the tough fibrous material in plant tissues), making nutrients available to the ants.

The ants excavate nests and build nice, cozy, safe chambers for their fungal gardens; they clean off debris and parasites from the fungi; and even produce special antibiotics in their bodies to ward off or kill infectious organisms that might attack their crops. In return, the fungi produce swellings at the tips of their hyphae (long strings of fungal cells) that are rich in proteins, sugars and other nutrients. The ants dine on these nutritious swellings, called gongylidia.

The fungi are ensured a copious food supply and a stable, nurturing environment in which to live. The fungi rely upon the ants for reproduction. Before new queen ants fly off to mate and found their own colonies, they tuck a bit of fungus in their mandibles to start their new gardens. The fungi growing in virtually every leaf-cutter garden are actually clones of the same fungus farmed by ants 25 million years ago. Fungi feed themselves quite ably, absorbing nutrients from organic materials. Algae and cyanobacteria are also adept at providing for their own nutritional needs by turning sunlight into energy through photosynthesis. Yet many thousands of years ago, some fungi merged with some algae (or cyanobacteria in some cases) to create a new kind of partnership called a lichen.

Some 20,000 different kinds of lichens live in such diverse habitats as the surfaces of rocks in arctic tundra and desert sands as well as the bark of trees in bayous and the sides of buildings. Bacteria are very small. Yet despite their size they show a surprising degree of structural complexity. In this section we will look at the various compounds that make up a microbe and how these are put together. Learning the structure of a microbe helps in understanding how a microbe functions. To drive this point home, here are a few examples.

Disease causing bacteria (pathogens) have various structures that enhance their ability to cause illness. One important property is the ability to attach to the intended victim. pili, a proteinaceous surface structure on bacteria, are critical in this process. Microbes are also capable of exchanging genetic information by mating. This process involves another type of surface structure, the F-pilus.

Bacteria will take steps to insure their survival. This can take the form of creating resting structures that allow the microbe to "sleep" during bad times. During abundant times, many microbes will store excess carbon, nitrogen, sulphur or phosphorous in inclusions in the cell. Not only is structure important to understand functional relationships, it's also fascinating to observe what these little architects come up with. There are some universal structures that all bacteria have. The basic building blocks of life, DNA, RNA, and protein, are common to all organisms not just microbes. Also, all microbes have a cell membrane. Much of what we know about these structures was obtained by studying bacteria, yet another reason to study them. Finally, most bacteria have a cell wall, but not all.

Whether measured by the number of organisms or by total mass, the vast majority of life on this planet is microscopic. These teaming multitudes profoundly influence the make-up and character of the environment we live in. Presently, we know very little about the microbes that live in the world around us because less than 2 % of them can be grown in the laboratory. Understanding which microbes are in each ecological niche and what they are doing there is critical for our understanding of the world. Microbes are the major actors in the synthesis and degradation of all sorts of important molecules in environments. Cyanobacteria and algae in the oceans are responsible for the majority of photosynthesis on earth. They are the ultimate source of food for most ocean creatures (including whales) and replenish the world's oxygen supply.

Cyanobacteria also use carbon dioxide to synthesize all of their biological molecules and thus remove it from the atmosphere. Since carbon dioxide is a major greenhouse gas, its removal by cyanobacteria affects the global carbon dioxide balance and may be an important mitigating factor in global warming. In all habitats, microorganisms make nutrients available for the future growth of other living things by degrading dead organisms. Microbes are also essential in treating the large volume of sewage and waste water produced by metropolitan areas, recycling it into clean water that can be safely discharged into the environment. Less helpfully (from the view of most humans) termites contain microorganisms in their guts that assist in the digestion of wood, allowing the termites to extract nutrients from what would otherwise be indigestible. Understanding of these systems helps us to manage them responsibly and as we learn more we will become ever more effective stewards.

The vast majority of reported cases of foodborne illness occur as individual or sporadic cases. In most cases these originate, and occur, in the home. An outbreak occurs when two or more people suffer foodborne illness after consuming food from a contaminated batch.

Often, a combination of events contributes to an outbreak, for example, food might be left at room temperature for many hours, allowing bacteria to multiply which is compounded by inadequate cooking which results in a failure to kill the dangerously elevated bacterial levels. Outbreaks are usually identified when those affected know each other. However, some are identified by public health staff from unexpected increases in laboratory results for certain strains of bacteria.

Energy is essential for our industrial society and microbes are important players in its production. A significant portion of natural gas comes from the past action of methanogens (methane-producing bacteria). Numerous bacteria are also capable of rapidly degrading oil in the presence of air and special precautions have to be taken during the drilling, transport and storage of oil to minimize their impact. In the future microbes may find utility in the direct production of energy. For example, many landfills and sewage treatment plants capture the methane produced by methanogens to power turbines that produce electricity. Excess grain, crop waste and animal waste can be used as nutrients for microbes that ferment this biomass into methanol or ethanol. These biofuels are presently added to gasoline and thus decreasing pollution. They may one-day power fuel cells in our cars, causing little pollution and having water as their only emission.

Microorganisms used in research have many useful properties. They will grow on simple, cheap medium and will often rise to large populations in a matter of 24 hours. It is easy to isolate their genomic material, manipulate it in the test tube and then place it back into the microbe. Due to their large populations it is possible to identify rare events and then with the use of powerful selective techniques isolate interesting bacterial cells and study them. These advantages have made it possible to rapidly test hypothesis in laboratory microorganisms that would have taken much longer in other organisms. Using microbes scientists have expanded our knowledge about life; here are a few examples.

In scientific research, microorganisms have been indispensable instruments for unlocking the secrets of life. The molecular basis of heredity and how this is expressed as proteins was described through work on microorganisms. Due to the similarity of life at the molecular level, this understanding has helped us to learn about all organisms, including ourselves. A beer is any variety of alcoholic beverages produced by the fermentation of starchy material derived from grains or other plant sources. The production of beer and some other alcoholic beverages is often called brewing. Historically, beer was known to the Sumerians, Egyptians, and Mesopotamians, and dates back at least as far as 4,000 BC. Because the ingredients used to make beer differ from place to place, beer characteristics (type, taste, and colour) vary widely.

Typically, beers are made from water, malted barley, hops, and fermented by yeast. The addition of other flavourings or sources of sugar is not uncommon. Because beer is composed mainly of water, the source of the water and its characteristics have an important effect on the character of the beer. Many beer styles were influenced or even determined by the characteristics of the water in the region.

Among malts, barley malt is the most often and widely used owing to its high enzyme content (which facilitates the breakdown of the starch into sugars) but other malted and unmalted grains are widely used, including wheat, rice, maize, oats, and rye.

Hops are a comparatively recent addition to beer. They contribute a bitterness that balances the sweetness of the malt and have a mild antibiotic effect that favours the activity of brewer's yeast over less desirable organisms. Enzymes in yeast, in a process called fermentation, metabolize the sugars extracted from the grains, producing many compounds including alcohol and carbon dioxide. Dozens of strains of natural or cultured yeasts are used by brewers, roughly sorted into three kinds: ale or top-fermenting, lager or bottom fermenting, and wild yeasts. Top-fermenting means that the yeast ferments in the top of the fermenting vessel. Conversely, bottom-fermenting means that the yeast ferments in the bottom of the fermenting vessel. The scientific name for ale yeast is Saccharomyces cerevisiae, an important model organism in molecular and cell biology; Saccharomyces carlsbergensis is the scientific name for lager yeast.

During the process of filtration (also called fining or clearing), some brewers add agents to beer that are not required to be published as ingredients. Since these finings may include animal extracts, vegans and others concerned with the use or consumption of animal products may wish to contact the brewer for specific details of the filtration process. Isinglass finings are a common animal-derived clarifying agent, extracted from fish. Alternatively, Irish moss is a commonly used plant-based clarifying agent. One pint of beer typically contains about two to three units of alcohol, although alcohol content can vary significantly with style and brewer.

Almost any sugar or starch-containing food can naturally undergo fermentation, and so it is likely that beer-like beverages were independently invented in cultures throughout the world. In Mesopotamia, the oldest evidence of beer is on a 6000-year old Sumerian tablet which shows people drinking a beverage through reed straws from a communal bowl. Beer is also mentioned in the Epic of Gilgamesh, and a 3900-year old Sumerian poem honoring the brewing goddess Ninkasi contains the oldest surviving beer recipe, describing the production of beer from barley via bread. Beer became vital to all the grain-growing civilizations of classical antiquity, especially in Egypt and Mesopotamia. Beer was important to early Romans, but during Republican times wine displaced beer as the preferred alcoholic beverage, and beer became considered a beverage fit only for barbarians. Tacitus wrote disparagingly of the beer brewed by the Germanic peoples of his day.

Most beers until relatively recent times were what we would now call ales. Lagers were discovered by accident in the sixteenth century when beer was stored in cool caverns for long periods; they have since largely outpaced ales in volume. The use of hops for bittering and preservation is a medieval addition. Hops were cultivated in France as early as the 800s. The oldest surviving written record of the use of hops in beer is in 1067 by Abbess Hildegard of Bingen: "If one intends to make beer from oats, it is prepared with hops." In 15th century England, an unhopped beer would have been known as an ale, while the use of hops would make it a beer. Hopped beer was imported to England (from the Netherlands) as early as 1400 in Winchester and hops were being planted on the island by 1428. The Brewers Company of London went so far as to state "no hops, herbs, or other like thing be put into any ale or liquore wherof ale shall be made — but only liquor (water), malt, and yeast." However, by the 16th century, "ale" had come to refer to any strong beer, and all ale and beer were hopped.

Methods of brewing changed very little from that time. In 1953, New Zealander Morton W Coutts developed the technique of continuous fermentation which was the first major change to brewing since the 16th century. Morton patented his process which revolutionized the industry by reducing a four-month long brewing process to less than 24 hours. His process is still used by many of the world's major breweries today, including Guinness.

In 1516, the duchy of Bavaria adopted the Reinheitsgebot, perhaps the oldest food regulation still being used. The Reinheitsgebot ordered that the ingredients of beer be restricted to water, barley, and hops. The law soon spread throughout Germany, and has since been updated to reflect modern trends in beer brewing. To this day, the Reinheitsgebot is (controversially) considered a mark of purity in beers.

Some prokaryotes are capable of growing under unimaginably harsh conditions and define the extreme limits of where life can exist. Some species have been found growing at near 100 °C in hot springs and well above that temperature near deep-sea ocean vents. Others make their living at near 0 °C in freshwater lakes that are buried under the ice of Antarctica. The ability of microbes to live under such extreme conditions is forcing scientists to rethink the requirements necessary to support life. Many now believe it is entirely possible that Jupiter's moon Europa may harbor living communities in waters deep below its icy crust. What may the rest of the universe hold?

Until recently, while we could study specific types of bacteria, we lacked a cohesive classification system, so that we could not readily predict the properties of one species based on the known properties of others. Visual appearance, which is the basis for classification of large organisms, simply does not work with many microbes because there are few distinguishing characteristics for comparison between species even under the microscope. However, analysis of their genetic material in the past 20 years has allowed such classification and spawned a revolution in our thinking about the evolution of bacteria and all other species. The emergence of a new system organizing life on Earth into three domains is attributable to this pioneering work with microorganisms.

The fruits of this basic research have been used by scientists to understand microbial activity and therefore to shape our modern world. Human proteins, especially hormones like insulin and human growth factor, are now produced in bacteria using genetic engineering. Our understanding of the immune system was developed using microbes as tools. Microorganisms also play a role in treating disease and keeping people healthy. Many of the drugs available to treat infectious disease originate from bacteria and fungi.

One last recent role of microbes in informing us about our world has been the tools they provide for molecular biology. Enzymes purified from bacterial strains are useful as tools to perform many types of analyses. Such analyses allow us to determine the complete genome sequence of almost any organism and manipulate that DNA is useful ways. We now know the entire sequence of the human genome, with the exception of regions of repetitive DNA, and this will hopefully lead to medical practices and treatments that improve health. We also know the entire genome sequence of many important pathogens and analysis of this data will eventually lead to an understanding of the function of critical enzymes in these microbes and the development of tailor-made drugs to stop them. The tools of molecular biology will also affect agriculture. The complete genome sequence of the plant Arabidopsis (a close relative of broccoli and cauliflower) is now known. This opens a new avenue to a better understanding of all plants and hopefully improvements in important crops.

Before we begin the adventure that we call learning microbiology (it can be thought of as an adventure!), a look at the history of microbiology will help you to understand the contributions of those who have come before. This perspective will hopefully give you an appreciation of their efforts and put the body of knowledge we will look at in the context of history. Keep in mind that microbiology is a relatively young science. It was only 130 years ago that it became possible to seriously study microorganisms in the laboratory with most of our understanding of microbes coming in the last 60 years.

The history of microbiology, like all human history, is not a catalog of linear progress, but is more of an interweaving of the careers of bright individuals and their insights. Each new discovery relied on previous ones and in turn spawned further inquiry. A web of interdependent concepts evolved over time through the work of scientists in many related disciplines and nations. Often the research of one individual impacted the efforts of another studying a completely different problem. Keep this in mind as you look at this history.

Below we present several journeys through this web, mentioning some individuals who were particularly important in the progression. This history reflects our view of important events of the past, but is by no means comprehensive. We will first look at the development of the techniques for handling microorganisms, since everything else in microbiology depends upon these procedures. Next, we will examine how these techniques helped to settle an old debate, the question of spontaneous generation. Then, we will look at the history of infectious disease. The science of microbiology had its most significant early impact on human health, discovering the cause of the major killers of the day, and then methods to treat them. As microbiology matured, scientists began to look at what non-pathogenic microbes were doing in the environment and we will look a bit at the history of general microbiology. An examination of the events that lead to the understanding of life at the molecular level and the profound impact this has had on microbiology and on society in general. By examining a sequence of discoveries we hope to give you a sense of how science builds on itself. Notice especially how ideas in one path influence ideas in another.

Koch was convinced that microbes caused some diseases. However, to test this idea, he needed to isolate the causative agent. Almost all samples from diseased animals or any natural surface contained many different microbes and it was impossible to tell which one was the problem. A method was needed to separate these different bacteria. The most common method of isolation was to continually dilute a sample in liquid broth in hopes that at high enough dilution, only one type of microbe would be found. A problem with this method is that only the most populous microbe would be isolated, but that might not be one causing the disease.

There were other technical problems as well with such a liquid-based system, so a solid medium would provide distinct advantages. Koch has tried gelatin for these experiments with unsatisfactory results. Building on the work of Brefeld and Schroeter, Koch used potato slices as a solid medium and observed that a boiled potato left in the open air would develop tiny circular raised spots. Examination of these spots revealed they were made up of microorganisms and each spot had just one type of microbe in it. He realized that these colonies were pure cultures of bacteria and probably arose from a single species of microbe from the air that landed on the potato. By boiling a potato, slicing it with a hot knife and keeping it in a sterile container with a lid, Koch could keep the potato sterile. But if a sample from a disease animal was smeared across the potato, colonies arose, each being pure isolates from the animal. By then testing these isolates in animals, Koch was able to isolate the cause of anthrax, Bacillus anthracis.

Potatoes failed to support the growth of many microorganisms and Koch and his laboratory were constantly frustrated by the lack of a good solid medium. Walter Hesse joined Koch's laboratory to do studies on air quality, showing a remarkable attention to detail and patience in his work. His wife, Angelina Fannie Hesse along with raising their three sons, also would assist her husband with his research in the laboratory. Walter was attempting to do his air quality experiments using medium containing gelatin as the solidifying agent. In the summertime temperatures would often rise above the melting point of gelatin. In addition, microbes would often grow in the cultures that were capable of degrading gelatin and in both cases this would cause liquefaction of the medium, ruining the experiments. One day the frustrated scientist asked Lina (as she was called) why her jellies and puddings stayed solid even in the hot summer temperatures. She told him about agar-agar, a heat resistant gelling agent that she had learned about while growing up in New York from a Dutch neighbour who had emigrated from Java. Development of the new agent by Angelina and Walter led to a resounding success. Few microbes are able to degrade agar and it melts at 100 °C yet remains molten at temperatures above 45°C. This allows the mixing of the agar with heat-sensitive nutrients and microbes. After solidification, it will not melt until a temperature of 100°C is again attained, allowing the easy cultivation of pathogens. It can also be stored for long periods of time allowing the cultivation of slow-growing microorganisms. Any type of broth can be mixed with agar and this gives great flexibility in the kinds of medium that can be made, thus many more types of microbes could be cultivated.

Koch laboratory also developed methods of pure culture maintenance and aseptic technique. Aseptic technique involves the manipulation of pure cultures in a manner that prevents their contamination by outside microorganisms and equally important prevents their spread into the environment. Remember that Koch was studying some of the most devastating microbial pathogens of the period and their release could potentially cause disease in the scientists working on them. These procedures were also absolutely critical because they allowed careful study of pure microorganisms, making it possible to identify the role of each microbe in a given situation.

Another problem in the cultivation of microbes was solved by Julius Petri while working in Koch's laboratory. Solid medium was poured on glass plates and allowed to spread and harden. Once cooled it allowed a solid surface for streaking. However, creation of these plates required great care since exposure to the air (and the microbes in it) often lead to contamination. In addition, to prevent contamination of plates during incubation, a cumbersome bell jar was used. If one wanted to view samples, the plate had to be removed from the jar, further exposing it to unwanted microbes in the air. In 1887 Petri developed a shallow glass dishes, with one having a slightly larger diameter than the other. Medium is poured into the smaller dish and the larger one serves as a cover. This simple device solved all of the above problems and has taken on the name of its inventor, the petri plate.

These same techniques are essential in studying all microorganisms. Collectively the above techniques have been used to isolate and identify thousands of different microorganisms. As a testament to the significance of their achievement, these techniques are practiced with remarkably little change in every laboratory that works with microorganisms today. The table below lists some of the early advances that helped to develop the practice of microbiology.

Spontaneous generation is the hypothesis that some vital force contained in or given to organic matter can create living organisms from inanimate objects. Spontaneous generation was a widely held belief throughout the middle ages and into the latter half of the 19th century. In fact, some people still believe in it today. The idea is attractive because it meshed nicely with the prevailing religious views of how God created the universe. There was a strong bias to legitimize the idea because this vital force was considered a strong proof of God's presence in the world. Many recipes and experiments were offered in proof. To create mice, a recipe called for dirty underwear and wheat grain be mixed in a bucket and left open outside. In 21 days or less, you would have mice. The real cause may seem obvious from a modern perspective, but to the proponents of this idea, the mice spontaneously arose from the wheat kernels.

Another often-used example was the generation of maggots from meat that was left in the open. The failing here was revealed by Francesco Redi in 1668 with a classic experiment. Redi suspected that flies landing on the meat laid eggs that eventually grew into maggots. To test this idea he used three pieces of meat. One piece of meat was placed under a piece of paper. The flies could not lay eggs onto the meat and no maggots developed. The second piece was left in the open air, resulting in maggots. In the final test a third piece of meat was overlaid with cheesecloth. The flies were able to lay the eggs into the cheesecloth and when this was removed no maggots developed. However, if the cheesecloth containing the eggs was placed on a fresh piece of meat, maggots developed, showing it was the eggs that "caused" flies and not spontaneous generation. This helped to end the debate about spontaneous generation for large organisms. However, spontaneous generation was so seductive a concept that even Redi believed it was possible in other circumstances.

The concept and the debate were revived in 1745 by the experiments of John Needham. It was known at the time that heat was lethal to living organisms. Needham theorized that if he took chicken broth and heated it, all living things in it would die. After heating some broth, he let a flask cool and sit at a constant temperature. The development of a thick turbid solution of microorganisms in the flask was strong proof to Needham of the existence of spontaneous generation. Lazzaro Spallanzani later repeated the experiments of

Needham, but removed air from the flask, suspecting that the air was providing a source of contamination. No growth occurred in Spallanzani's flasks and he took this as evidence that Needham was wrong. Proponents of spontaneous generation discounted the experiment by asserting that air was required for the vital force to work.

It was long suspected that living things were the agents of disease. In volume 6 of his epic poem De Rerum NaturaI (On the Nature of the Universe) written sometime around 50 B.C., Titus Lucretius Carus speculates about invisible atoms causing disease. This was only one idea among many and some thought that an imbalance in humors caused illness, while others felt that supernatural forces were at work. The prevailing theory held by most doctors of the 19th century was that chemical toxins were carried from an ill patient to others, causing them to contract the same malady.

The bacteria that were known to be present were seen as a symptom of the disease and not its cause. Ignaz Semmelweis, a Hungarian physician working in Vienna, made the first breakthrough in the true nature of disease. He realized that asepsis in obstetrical wards could prevent the transmission of childbirth fever from patient to patient and instigated a policy for all attending physicians to wash their hands with chloride of lime (a mixture of calcium chloride hypochlorite, CaCl(OCl); calcium hypochlorite, $Ca(OCl)_2$; and calcium chloride, $CaCl_2$) between patients.

This innovation dropped the mortality rate or mortality from 18% to 2.4%. Ignaz became a vigorous proponent of these ideas, but the Hungarian doctor's efforts were opposed by many who could not accept that physicians themselves could be responsible for spreading bacterial infection. Ridicule of his idea caused him to move from Vienna to Pest, Hungary and ultimately played a role in a nervous breakdown. Ironically Semmelweis died from an infection that he contracted during a surgery he performed, while recovering from his nervous breakdown. Before his death he published his ideas in a paper The Cause, Concept, and Prophylaxis of Childbed Fever in 1861. The work was then ignored for 17 years, which raises an interesting point about the culture of science. Radical ideas, even those that are correct and can save lives, are sometimes ignored. It takes time to overcome the dogma of the day. The personalities involved and the negative light it might throw on past practice play a large role in the rate of acceptance of a new idea. His book was also very confusing dues to its poor writing and this also contributed to the obscurity of his ideas.

1. The specific organism should be shown to be present in all cases of animals suffering from a specific disease, but should not be found in healthy animals.
2. The specific microorganism should be isolated from the diseased animal and grown in pure culture on artificial laboratory media.
3. This freshly isolated microorganism, when inoculated into a healthy nonimmune laboratory animal, should cause the same disease seen in the original animal.
4. The microorganism should be reisolated in pure culture from the experimental infection.

This is his most famous contribution to science and it is a testament to the utility of these postulates that they are stilled used today to discover the cause of new emerging diseases. Koch went on to apply these principles in the study of many other diseases including tuberculosis, cholera and sleeping sickness. It should be pointed out that Koch's postulates cannot be applied to all diseases. For example if a disease-causing microbe has humans as its sole host and has a significant possibility of causing death, it would be unethical to apply this microbe to test humans as dictated by postulate 3. Also, it is not always possible to obtain a disease-causing microbe in pure culture.

Koch developed the tools for obtaining pure cultures to attack the problem of disease. Advances in science often come from innovations in the available technology. Robert Koch was an important microbiologist because his pioneering work in the isolation and characterization of bacterial diseases helped to identify the causes of many of the maladies plaguing humanity. Further work by other scientists then began the long road to conquering them.

Smallpox was a feared disease throughout human history and justifiably so. It was highly contagious and almost everyone eventually became infected. Mortality rates were as high as 25% in adults and closer to 40% in children. Those who did survive often had scarring due to the blister-like pustules that form on the skin, but they obtained life-long immunity to the disease. As far back at the 11th century in India and China it was realized that liquid from the pustules of a smallpox victim, when scratched on the skin of a healthy patient, would most often cause mild disease. This intentional infection, termed variolation, would also give life-long protection against the virus. Lady Mary Wortley Montgue, wife of ambassador to the Ottoman Empire, introduced variolation to England in 1721 and it became a popular practice throughout Europe. Washington even began variolating the Continental Army in 1776.

Variolation had some deleterious side effects. Serious skin lesions inevitably resulted at the site of inoculation, often accompanied by a generalized rash or even a full case of smallpox. The fatality rate from variolation was 1 to 2 %. Today we would find this level of fatality to be unacceptable, but this risk still represented a significant advance. In the 1796, Edward Jenner, an English country physician, went in search of a more predictable and safer method of protection against the disease. He noticed that milkmaids rarely contracted smallpox. Further investigation revealed they often contracted cowpox from their charges. In a perceptive leap of faith, Jenner hypothesized that cowpox was related to smallpox and contraction of the former would protect against the latter. In a classic experiment (and one that would land you in jail today), Jenner inoculated a young patient with cowpox and later challenged him with smallpox. The boy did not become ill and Jenner was responsible for the creation of a safer method of protection against small pox. It is important to stress that the nature of these diseases and their viruses would not be known for over 100 years. Jenner was ahead of his time.

Beginning around 1876 Pasteur's studies on chicken cholera led to the development of vaccines to fight the disease. Cholera was a serious problem since it was able to spread through a barnyard and wipe out a flock in as little as 3 days. It was transmitted by contaminated food or animal excrement. Pasteur had identified the cholera bacillus and was growing it in pure culture. When injected with it, a chicken invariably died within 48 hours.

Then, as often happens in scientific research, luck intervened. During the heat of the summer, Pasteur returned to Paris and left the cholera cultures used for infection stored on the shelves of his laboratory in Arbois, France. Upon returning, something had happened to the cultures, they no longer caused disease when tested in chickens. With some impatience for the time they were wasting, his group set to work making new cultures of the bacillus and tested these batches on both new birds and also those previously inoculated with the ineffective strain. To their amazement the previously injected birds were unaffected by the fresh bacillus culture, while the new birds all died. Pasteur immediately realized that this was similar to the studies of Jenner. Pasteur then developed a method for creating cultures that would confer immunity, but not cause disease. In honour of Jenner's accomplishments Pasteur coined the term vaccination (vacca = cow in Latin) for the process of immunization against disease. In this and several of Pasteur's other discoveries, luck played a part, but it was only helpful because he tenaciously pursued "odd" results and had the insight to arrive at important conclusions. In Pasteur's famous words, " In the field of observation, chance favours only the prepared.

We now pick up another thread though the web of change that began with the work of Paul Ehrlich. By 1885 it was becoming clear that the causative agents of many illnesses were microorganisms. As scientists manipulated these microbes in the lab they found that certain dyes and other compounds were able to inhibit their growth. This inspired Ehrlich to propose that chemicals may exist that will kill the microbe, but not the patient. The hope was that by having a patient take these chemicals they could cure the illness. One of the diseases he hoped to cure was syphilis, which had reached epidemic proportions in Europe. Little did he realize it would be a 17-year odyssey before he would develop salvarsan, the first effective chemotherapeutic agent. Salvarsan was the 606th chemical he tried and is an arsenic compound that effectively kills Treponema pallidum, the causative agent of syphilis. However, the treatment had many problems, causing long lasting health complications for those individuals who used it.

In addition, despite Treponema being quite sensitive to salvarsan, the physician had to administer it intravenously for optimum effectiveness. This type of injection was a recent development and many doctors were leery of trying the procedure. In London a young physician by the name of Alexander Fleming, then in the Army Medical Corps, was one of the few that was willing to treat patients, even getting the nickname private 606 from his burgeoning practice. His work validated the effectiveness of salvarsan against syphilis and convinced others to try the treatment. Fleming was a physician by training, but spent most of his time studying bacteria, and his success with salvarsan motivated him to search for other antibacterial agents. His first discovery was lysozyme, an enzyme produced by many organisms, including humans, that lysed some bacteria. This enzyme is not useful as a therapeutic agent because it is difficult to administer as a drug, but Fleming did develop titration methods and assays that would become very useful. In September of 1928, before leaving on a summer holiday, Fleming streaked some plates of Staphylococcus aureus and left them to incubate until his return.

In an improbable set of circumstances, the beginning of the holiday was cold, allowing some contaminating mold spores (that had blown in from a nearby window) to grow up on some of the plates. The temperature then increased encouraging the growth of the Staphyloccoccus. Many experimenters when confronted with a contaminated plate will look for the trash bin, but Fleming instead spent some time examining it. The fungus had a zone of clearing around it where the Staphylococcus colonies would not grow, suggesting the fungus was producing an antibacterial compound that had diffused into the medium. Intrigued, he cultured the fungus, a Penicillium mold, and eventually isolated a soluble extract that could kill bacteria and treat localized infection. He called the new compound penicillin after the mold it came from. Due to the technology available, it was very difficult to prepare a solution that could be used throughout the body without causing problems. World War II added a greater urgency to the search for compounds that could fight infectious disease. Wounded soldiers, if they survived the initial injury, would often develop life-threatening infections and there were no effective drugs to combat them. In 1939 Howard Florey and Ernst Chain began a systematic study of antimicrobial compounds in hopes of developing treatments for these soldiers and ran across Fleming's report written 9 years earlier. They now were able to purify the compound completely and describe its high potency against microbes. The availability of penicillin during World War II saved countless lives. The rediscovery of penicillin touched off a search for other microbes producing substances that could kill or inhibit microbes, leading to the discovery of many more antimicrobials. With Florey and Chain, Fleming was awarded the Noble prize in Medicine and Physiology in 1945.

Martinus Beijerinck was originally trained as a botonist and began his work studying the microorganisms that were present in and around plants. He soon began experiments with microbes in the soil. His greatest contribution was the development of enrichment media. Previously, microbes were cultivated on medium consisting of potatoes or extracts of left-over animal renderings. Such media would support the growth of many different bacteria, with chance and population density dictating what became dominant in the culture. In many cases this was exactly what was desired, but in other cases it was of interest to find bacteria capable of performing certain chemical conversions.

Beijerinck discovered that by adding or removing certain compounds from the medium or incubating under different conditions, it was possible to favor the growth of certain microbes and prevent the growth of others. An example is Beijerinck's work with nitrogen-fixing bacteria, which are important in agriculture and the global cycling of nitrogen. They are capable of taking nitrogen gas (N2) from the air and reducing it to ammonia (NH3). This is an important property since reduced nitrogen compounds such as ammonia are the only compounds most organisms, including agriculturally important plants, are able to use as a nitrogen source. Beijerinck wanted to isolate an organism capable of fixing nitrogen in the presence of air, because all previous isolates fixed nitrogen only under anaerobic conditions. By making medium that did not contain a source of fixed nitrogen and then incubating in the presence of air (containing N2), he demanded that any microbe growing in the medium had to be able to derive its nitrogen by performing aerobic nitrogen fixation. By this method he succeeded in isolating a new microbe (Azotobacter chroococum) with these capabilities. Similar selective culture techniques (as he liked to call them) enabled his laboratory to isolate sulphur-reducing and sulphur-oxidizing bacteria, Lactobacillus species, green algae and many other microbes. Sergei Winogradsky was also interested in soil bacteria, especially those involved in the cycling of nitrogen and sulphur compounds. He was one of the first to isolate microorganisms responsible for the conversion of these elements in the soil, obtaining pure cultures of bacteria capable of the conversion of ammonia to nitrate by microorganisms in the soil. Winogradsky also studied the consumption of hydrogen sulfide gas by sulphur-oxidizing bacteria directly in their natural habitat.

When working with Beggiatoa (one of these sulphur-oxidizing bacteria) he discovered it was obtaining its cell carbon from carbon dioxide. He used the term autotrophy to describe this property, a radical idea at the time. This led Winogradsky to propose the concept of chemolithotrophy-bacteria capable of growth using purely inorganic sources of carbon and energy. He realized the impact these processes have on our environment long before others accepted the idea.

As a result of the work of Winogradsky and Beijerinck there was great enthusiasm for identifying and classifying the bacteria inhabiting our natural world. For the first part of the 20th century many scientists isolated microbes and made proposals for their organization into genera and species. In 1909 Sigurd Orla-Jensen suggested a classification scheme based on the functions and abilities of the bacteria, such as growth on certain compounds or the production of specific by-products. The Society of American Bacteriologists, later to become the American Society of Microbiology, applied this technique to prepare a report on the classification of bacteria. This eventually evolved into Bergey's Manual of Determinative Bacteriology in 1923 and subsequent editions have become authoritative reference works. There was great optimism that as more bacteria were brought into pure culture a clear organization, based on the evolutionary history of microbes, would emerge from studies of their physiology. However, y the 1940's it became clear that bacteria were unwilling to go along with this idea because classification based on one set of properties was often inconsistent with classification using a different set of properties. Scientists threw up their hands and gave up the idea of finding a system of classification based on growth and morphological characteristics.

Several decades passed during which bacteria became tools for understanding life, but their place in evolution was of little interest. Then in 1961 McCarthy and Bolton developed methods for the comparison of genetic material between species and Pauling and Zuckerkandl formalized the idea of using the make-up of biological molecules, their sequence, as a way to determine phylogenetic relationships. Later, Carl Woese began to use this insight, choosing the sequence of the 16S rRNA from the ribosomes of bacteria as basis for evolutionary comparisons. This analysis was a watershed event for the evolutionary classification of bacteria and in 1977 Woese used this technique to assert that known bacteria should be divided into two separate domains that we now call Bacteria and Archaea, radically changing our view of the microbial world. This hypothesis, termed endosymbiosis, has been strongly supported by a range of subsequent biochemical data. In subsequent years, determination of 16S rRNA sequences, and then complete genomes, of microorganisms has enabled the natural classification of microorganisms. A review by Olsen, Woese and Overbeek provided a catalyzing summary of the tree of life for microbes, including eukaryotic microorganisms. This reorganization has reshaped our view of the tree of life for all organisms and helped to reassert the important role of microbes in shaping the past and affecting the future of the earth. A renewed interest in environmental microbiology, using new molecular tools based often on 16S rRNA, is revealing much more about the microbes present in our world and their influence on it.

In the early part of the 20th century techniques were developed to examine the inner workings of the cell and much of the work was performed in bacteria due to their experimental accessibility. Before this period, the method of turning genetic information into the proteins that carry out cellular processes was completely unknown. Indeed, the very nature of genetic material was being hotly debated when in 1928 Fred Griffith discovered transformation in bacteria.

Griffith knew that when Streptococcus pneumoniae are injected into mice, they cause rapid deterioration and death. Griffith isolated strains of the bacteria that were no longer capable of producing an outer slime layer and appeared rough when grown on solid medium in contrast to the smooth colonies of the original isolate. When these rough strains were injected into mice, no illness resulted. Similarly, when a smooth microbe was heat-killed and then injected, the mice showed no signs of infection. A surprise was waiting for Griffith when he mixed the dead cells of a smooth isolate with live cells of a rough isolate and then injected this into mice; they died.

When bacteria were isolated from these dead mice, they were all smooth. Griffith hypothesized that the ability to create the slime layer was passed from the dead smooth cells to the viable rough cells, making them pathogenic again. The process became known as transformation. He was ridiculed, as most scientists believed his preparation were contaminated with viable smooth cells. It was not until 1944 that his student Oswald Avery and coworkers repeated the experiments of Griffith, reproducing his results and discovering that DNA was the material from the dead smooth cells that transformed the rough mutant. This was strong proof that the hereditary material in cells was DNA, though the actual structure of DNA was not to be discovered for years. Further experiments in the 1940's clearly established this observation with Joshua Lederberg discovering two more ways that DNA could be transferred between bacterial cells, conjugation and transduction.

Peptones from autohydrolysed fish viscera for nisin and pediocin production; Vazquez JA et al.; Various peptones obtained from hydrolysed visceral homogenates of four fishery residues showed their suitability for promoting the growth of lactic acid bacteria, micro-organisms with particularly complex requirements regarding peptidic nutrients. The assay of several treatments with two bacterial species, producers of the two main bacteriocins (nisin and pediocin) demostrated that optimum conditions only imply a brief autohydrolysis at natural pH and room temperature, with subsequent steam-flow stabilisation. Later kinetic analysis of the cultures of both bacteria in the best media provided parameters which, for production of both biomass and bacteriocins (the latter behaved in the majority of cases as a mixed metabolite), indicate comparable or superior results to those found in costly commercial media, specifically recommended for culture of lactic acid bacteria. Lactic acid bacteria used in inoculants for silage as probiotics for ruminants; Weinberg ZG et al.; Many studies have shown the beneficial effects on ruminant performance of feeding them with silages inoculated with lactic acid bacteria (LAB). These benefits might derive from probiotic effects. The purpose of the current study was to determine whether LAB included in inoculants for silage can survive in rumen fluid (RF), as the first step in studying their probiotic effects. Experiments were conducted in the United States and Israel with clarified (CRF) and strained RF (SRF) that were inoculated at 10(6)-10(8) microorganisms/mL with and without glucose at 5 g/L. RF with no inoculants served as control. Ten commercial inoculants were used. The RF was incubated at 39 degrees C and sampled in duplicates at 6, 12, 24, 48, 72, and 96 h for pH and LAB counts. The results indicate that with glucose the pH of the RF decreased during the incubation period. In the SRF, the pH of the inoculated samples was higher than that of the controls in most cases. This might be a clue to the mechanism by which LAB elicit the enhancement in animal performance. LAB counts revealed that the inoculants survived in the RF during the incubation period. The addition of glucose resulted in higher LAB counts.

2

Technological Principles of Pasteurization

Physical and chemical factors which influence pasteurization process are the following:

a. temperature and time;

b. acidity of the products;

c. air remaining in containers.

Pasteurization processes. In pasteurising certain acid juices for example, there are two categories of processes:

Low pasteurization where pasteurization time is in the order of minutes and related to the temperature used; two typical temperature/time combinations are as following:

63° C to 65° C over 30 minutes or

75° C over 8 to 10 minutes.

Pasteurization temperature and time will vary according to:

- nature of product; initial degree of contamination;
- pasteurized product storage conditions and shelf life required.

In this first category of pasteurization processes it is possible to define three phases:

- heating to a fixed temperature;
- maintaining this temperature over the established time period (= pasteurization time);
- cooling the pasteurized products: natural (slow) or forced cooling.

Rapid, high or flash pasteurization is characterized by a pasteurization time in the order of seconds and temperatures of about 85° to 90° C or more, depending on holding time. Typical temperature/ time combinations are as follows:

88° C (190° F) for 1 minute;

100° C for 12 seconds;

121°C for 2 seconds.

While bacterial destruction is very nearly equivalent in low and in high pasteurization processes, the 121° C/2 seconds treatment give the best quality products in respect of flavour and vitamin retention. Such short holding times, however, require special equipment which is more difficult to design and generally is more expensive than the 63–65° C/30 minutes type of processing equipment. In flash pasteurization the product is heated up rapidly to pasteurization temperature, maintained at this temperature for the required time, then rapidly cooled down to the temperature for filling, which will be performed in aseptic conditions in sterile receptacles. Taking into account the short time and rapid performance of this operation, flash pasteurization can only be achieved in continuous process, using heat exchangers. Industrial applications of pasteurization process are mainly used as a means of preservation for fruits and vegetable juices and specially for tomato juice.

The thermopenetration problem is extremely important, especially in the case of the pasteurization of products packed in glass containers because it is the determining factor for the success of the whole operation. During pasteurization it is necessary that a sufficient heat quantity is transferred through the receptacle walls; this is in order that the product temperature rises sufficiently to be lethal to micro-organisms throughout the product mass. The most suitable and practical method to speed up thermopenetration is the movement of receptacles during the pasteurization process. Rapid rotation of receptacles around their axis is an efficient means to accelerate heat transfer, because this has the effect, among others of rapidly mixing the contents.

The critical speed of for this movement is generally about 70 rotations per minute (RPM). This enables a more uniform heating of products, reducing heating time and organoleptic degradation. Heating may precede or follow packaging. These principles of different temperature time combinations very largely determine the design parameters for heat preservation equipment and commercial practices.

The food processor will employ no less than that heat treatment which gives the necessary degree of micro-organism destruction. This is further ensured by periodic inspection by local sanitary authorities or by the importing countries sanitary services. However, the food processor also will want to use the mildest effective heat treatment to ensure highest food quality.

It is convenient to separate heat preservation practices into two broad categories: one involves heating of foods in their final containers, the other employs heat prior to packaging. The latter category includes methods that are inherently less damaging to food quality, where the food can be readily subdivided (such as liquids) for rapid heat exchange. However, these methods then require packaging under aseptic or nearly aseptic conditions to prevent or at least minimise recontamination.

On the other hand, heating within the package frequently is less costly and produces quite acceptable quality with the majority of foods and most of our present canned food supply is heated in the package. In practice, therefore, most of the canned food produced locally in developing countries should be heated within the package.

Food Irradiation

Food irradiation is one of the food processing technologies available to the food industry to control organisms that cause food-borne diseases and to reduce food losses due to spoilage and deterioration. Food irradiation technology offers some advantages over conventional processes. Each application should be evaluated on its own merit as to whether irradiation provides a technical and economical solution that is better than traditional processing methods.

Table: *Possible causes of spoilage (real or apparent) in canned goods*

Type of food: Acid and high acid foods (canned fruits)	
Condition of can	***Action to be taken to identify cause***
Insufficient vacuum or	Check vacuum and headspace in relation to storage
headspace	temperature and altitude
"Springer" or "flipper"	Cool can to 15°C and check if still domed. Check can for denting, if possible measure headspace volume change brought about by dents, by comparing can volume with volume of a sound can

Contd...

Condition of can	***Action to be taken to identify cause***
Hydrogen swell	Check degree of detinning in can especially at the liquid level. Look for scratches or pinholes in lacquer or tin coating. Check if can is still domed on cooling to 15°C.
"Hard" or "soft swell"	Leaker spoilage. Check can for gross seam faults, perforation due to corrosion or damage to seams. Examine contents for signs of spoilage and can interior for detinning at air/product interface.

Source: FAO/WFP, 1970

Applications

For products where irradiation is permitted, commercial applications depend on a number of factors including the demand for the benefits provided, competitiveness with alternative processes and the willingness of consumers to buy irradiated food products. There are a number of applications of food irradiation. For each application it is important to determine the optimum dosage range required to achieve the desired effect.

Too high a dosage can produce undesirable changes in texture, colour and taste of foods. Shelf-life extension. Irradiation can extend the shelf-life of foods in a number of ways. By reducing the number of spoilage organisms (bacteria, mould, fungi), irradiation can lengthen the shelf life of fruits and vegetables. Since ionising radiation interferes with cell division, it can be used as an alternative to chemicals to inhibit sprouting and thereby extend the shelf life of potatoes, onions and garlic.

Exposure of fruits and vegetables to ionising radiation slows their rate of ripening. Strawberries, for example, have been found to be suitable for irradiation. Their shelf-life can be extended three-fold, from 5 to 15 days.

Disinfestation. Ionising radiation can also be used as an alternative to chemical fumigants for disinfestation of grains, spices, fruits and vegetables. Many countries prohibit the importation of products suspected of being contaminated with live insects to protect the importing country's agricultural base. With the banning of certain chemical fumigants, irradiation has the potential to facilitate the international shipment of food products.

Global Developments

Consensus on Wholesomeness

In 1980, an FAD/IAEA/WHO Expert Committee reviewed in detail all the accumulated data on food irradiation from the past 40 years. The Expert Committee concluded that irradiation to an overall dose of 10 kGy (kilograys) presents no toxicological hazard and introduces no special nutritional or microbiological problems, thus establishing the wholesomeness of irradiated foods up to an overall average absorbed dose of 10 kGy.

Data were insufficient to formulate conclusions on applications of food irradiation above 10 kGy. Data on radiation chemistry, nutritional and microbiological aspects of food treated above 10 kGy is currently being compiled.

In 1983, the Codex Alimentarius Commission, an international group that develops global food standards for the FAO and the WHO, incorporated the 1980 Expert Committee's conclusions regarding the wholesomeness of irradiated foods into the Codex General Standard for Irradiated Foods. This proposed international standard was submitted to member countries to accept or to modify according to individual country needs. Currently most countries that allow food irradiation approve its use on a case-by-case basis.

The Codex Alimentarius Commission has also adopted a Recommended International Code of Practice for the Operation of Radiation Facilities for the Treatment of Foods. It is intended to serve as a guide for irradiator operators and government regulators.

International Trade

More than 30 countries have given clearances for the use of food irradiation to process some 40 food items and approximately 30 facilities world-wide treat food by irradiation processing. Approvals for additional items are being considered in many countries and many food irradiation facilities are being planned. It was anticipated in 1988 that by 1990 there could be approximately 50 commercial/demonstration irradiators in 25 countries.

Auxiliary Raw Materials

Auxiliary raw materials used in fruit and vegetable processing technologies play a major role in the determination of their physical and chemical characteristics, sensory properties and nutritive value.

Water

Water is one of the essential factors in the activity of the processing centres; according to the final utilisation, water can be classified in three categories:

a. for technological utilisation (when it comes into direct contact with raw materials and enters in the finished product's composition),

b. for steam generators and

c. for receptacle cooling, washing of equipment and general hygiene.

Water for Technological Uses

Water coming into direct contact with the raw materials used for processing (washing, blanching, etc.) or that used as filling liquid of some canned products, must be of drinking water quality in terms of its physico-chemical and microbiological conditions. More important even than fulfilling drinking water standards, water used for these purposes pests certain specific characteristics related to the technological step or the raw material treated during the processing.

When very hard water is used for blanching vegetables some pecto-calcium and pectomagnesium complexes are formed which starts the hardening of vegetable tissues. This process continues over the pasteurization of the finished product. When fruit is processed in sugar syrup, the use of hard water for the syrup preparation could induce the formation of a pectin-sugar-acid gel facilitated by the medium pH and presence of calcium salts.

Soft water has negative consequences associated with mineral and hydrosoluble substances and losses during blanching of vegetables. For some specific products such as peeled tomatoes, green beans and fine texture fruit, the addition of calcium salts (chlorure, sulphate, etc.) is employed to correct for low texture.

The water hardness is an essential factor when used as filling liquid for canned products; ideally, the hardness of the water should be adapted to the raw material species used for canning. Thus a hardness of 3° is good for beans, 5° to 9° for green peas, green beans, and for fruit and vegetables with a tendency to disintegrate should use even harder water. In the technological process of cucumber and gherkin preservation by natural acidification (lactic fermentation), water hardness has a paramount role. The literature maintains that as far as texture is concerned, the best results are obtained by using very hard water (about 30°); but since high magnesium and salt content has a negative effect on taste, in practice it is recommended that a water hardness of about 10° be used, which gives satisfactory results from both points of view.

Oxygen present in water can act as a corrosion factor in metal receptacles but this negative influence can be eliminated by preliminary boiling. An important factor is pH. Water for canning must be neutral or slightly basic. Acid water plays a major role in corrosion which is evident both on receptacles and on iron or copper equipment, where changes of product colour will be induced. More dangerous is the attack on lead pipes or to the mix used for can sticking and this can render the product toxic. For these reasons an acid water must be neutralised before use.

Water for Steam Generators

Two main conditions must be fulfilled:

- hardness has to be as low as possible, even zero, because precipitation of calcium salts can lead to the formation of encrustations ("crusts") in pipes and on equipment walls. For this reason, water treatment is practiced in the majority of installations;
- from a bacteriological point of view, the iron-bacteria must be eliminated with biological filters or oxidising substances. This is necessary in order to avoid iron hydroxide formed during equipment exploitation to deposit on the inner walls of the pipes.

The elimination of iron-bacteria is also of importance for the water used in processing steps.

Water for Receptacle Cooling and General Hygiene

This should be of drinking water standard. Where this is not available in sufficient quantity, the use of industrial water is acceptable but only for cleaning of production rooms/workshops.

Sweeteners

Sugar

Sugar is the conventional name applied to sucrose. Physically there is icing, granulated and lump sugar. In fruit and vegetable processing, sugar is used only in its granulated form; this quality must be in the form of uniform crystals, white, shining and completely soluble in water.

Concentration of various sugar solutions can be rapidly measured by refractometer reading or with areometers graduated in various ways: Brix, Baumé, etc. Sugar solubility in water is dependent upon temperature; for example, in order to obtain a saturated solution, one must dissolve 2040 g in one litre of water at 20° C and 4870 g at 100° C. Taking into account this temperature related solubility, in practice the majority of sugar solutions are prepared by heating the water. Water should be as soft as possible because the calcium salts can precipitate on boiling.

Corn Syrup (Liquid Glucose)

Corn syrup is obtained industrially by acid or enzymatic starch hydrolysis, using as starting raw materials maize (corn) or potatoes. In fruit processing, mainly in the production of marmalades, it is possible to use corn syrup. The average composition of this corn syrup is of about 32-40% dextrose (glucose), about 40% dextrins and 18-20% moisture. Sweetening power is 50% compared with sucrose.

In a 10%-20% proportion with sucrose, addition of corn syrup has certain advantages:

a. it improves the shine and texture of marmalade;
b. it prevents "sugaring" defect and
c. it reduces the too sweet taste of finished products obtained with sugar alone.

Salt

Salt is used in order to give to the finished products a specifically salty taste and as a preserving substance. From a chemical point of view the term salt means sodium chloride but in practice the product is never in a pure state. The presence of a significant quantity of magnesium chloride increases the hygroscopicity, gives a bitter taste and can induce corrosion of receptacles.

Food Acids

Acetic acid is in use as solutions of various concentrations which are known under the generic name of vinegar. Vinegar can be obtained:

a) from wine, alcohol, cider, beer, etc. by fermentation;
b) by dilution of acetic acid obtained by dry wood distillation or by synthesis.

From a quality point of view, wine vinegar is preferred, as it has a more pleasant flavour. In order to improve taste, other vinegar types are usually flavoured with spices.

In addition to its spicing and flavouring role, vinegar is used and acts as a preservation agent for some vegetables: cucumbers, acidified vegetables, etc.

- Citric acid
- Tartric acid
- Lactic acid.

Pectic Preparations

In fruit processing there many preparations and mixes known as "pectin" are used as liquid or powder extracts. From a practical point of view, the pectic preparations are classified as:

- strong pectins obtained from apples or citrus fruit peel; this category gives gels rich in sugar;
- weak pectins which gives gels with low proportion of sugar or even without sugar but with the addition of calcium salts.

The fruit industry uses mainly strong pectins. These preparations are characterised, from a commercial point of view, by the capacity of gelification, expressed in degrees. The degree of gelification represents the quantity of sugar in grams able to be transformed in a standard gel (65% sugar and pH=3) by 1 g pectin.

Intensive Sweeteners

"Calorie-reduced" and "low-calorie foods are widely used and are cornering an increasing share of the market. Sweeteners are making an important contribution to the manufacture of sweet foods in these categories. They make it possible to manufacture sweet products without "sweet" being necessarily synonymous with "high-calorie".

Diabetics need to restrict their intake of sugar and various carbohydrates similar to sugar or avoid them altogether. Sweeteners enable diabetics to enjoy sweet tastes without changing their lifestyle. Sweeteners do not contribute to the development of tooth decay; they do not degrade in the mouth to form the acids which are responsible for caries. Thus sweeteners offer consumers a number of advantageous and favourable properties above and beyond merely reducing calories.

Sunett is the trade mark of Hoechst AG for its high intensity sweetener acesulfame K. As an ingredient, it can be used for sweetening all foods produced industrially or at home, or to produce tabletop sweeteners. Like all other sweeteners, Sunett tastes more intensely sweet than sugar; it is about 200 times sweeter than sugar compared with a dilute stock solution.

Synergism

Sunett is notable for its pronounced synergism with other sweeteners. The synergistic effect leads as quantitative synergism to an intensification of the overall sweetness and as qualitative synergism to an improved taste. The synergism results in a marked intensification of the sweet taste of the blends, which can amount to 30-50% at usual concentrations. The favourable properties of Sunett, particularly its synergistic behaviour, can be used advantageously in the Sunett-Multi-Sweetener Concept. For high sweetness levels, blends of Sunett with other sweeteners, e.g. aspartame, are particularly favourable in many applications. The synergistic effect of the blend results in a taste that is particularly pleasant and rounded.

Alone or in blends with other sweeteners, Sunett is used mainly where only sweetness is important and no other properties of the foods are to be affected. This applies particularly to beverages (fruit syrups and juices, carbonated beverages, etc.). In a number of food products, sweet carbohydrates do not only provide the sweet taste; they have other functions to fulfil; for example, they act as bulking or texturing agents and as preservatives by reducing the water activity. In these types of products, Sunett and other sweeteners cannot be used on their own. They must be combined with other substances which perform the required functions. These may be bulking agents or sugar substitutes (for example polyols: sorbitol, mannitol, xylitol). Sunett can be combined with both groups of substances.

Uses. In carbonated soft drinks blends of Sunett with other sweet-tasting substances are often recommended. Beverages formulated using the Sunett Multi-Sweetener Concept, i.e. blends of Sunett with other sweeteners, are usually preferred because of their particularly well balanced and rounded sweetness profile. The taste of these blends is often superiors to single sweeteners.

Sunett is compatible with sugar and other sweet-tasting carbohydrates, both technically and in terms of taste. It can therefore be used in the production of soft drinks with a reduced sugar content. Drinks based on fructose and Sunett are suitable for diabetics. Soft drinks with a reduced sugar content show an excellent taste quality when Sunett or Sunett blends are added to bring them up to the usual level of sweetness.

Fruit nectars and fruit juice drinks differ from most carbonated drinks in that they contain a noticeable amount of sweet carbohydrates provided by the fruit juice. The amount of carbohydrate may vary depending on the type of fruit and amount of juice used. Amounts of up to 200 mg/l Sunett are often adequate as a single sweetener for the popular types of fruit nectars. With blends of Sunett and aspartame quantities in the order of 100150 mg/l Sunett and 50 mg/l aspartame or 60-70 mg/l of both Sunett and aspartame are sufficient.

Sunett offers such excellent stability that end products, such as drinks, show no reduction in sweetness performance during normal processing methods and storage periods. Sunett can withstand pasteurization, hot filling and aseptic filling without any loss of sweetness. Jams and marmalades. Sugar contributes a great deal to the texture and stability of conventional jams and marmalades. For the production of sugar-free products either the sugar must be replaced by comparable amounts of sugar substitutes or some of the functions of the sugar must be taken only by other components, such as suitable gelling agents.

Sugar-free jams and marmalades containing sweeteners are more susceptible to microorganisms than sugar-containing products. The risk of spoilage due to yeast fermentation or moulding can be prevented by pasteurization. However, this is only feasible for small jars which will be quickly consumed once they are opened.

In all other cases, it is advisable to add 0.05-0.1% potassium sorbate as a preservative, wherever this is permitted under the relevant food regulations. For sugar-free jams and marmalades, concentrations in the range of 500-2000 mg Sunett/kg of the finished product are appropriate. It is advisable to add Sunett in the form of an aqueous stock solution towards the end of the boiling process. Care must be taken to ensure that Sunett is evenly dispersed throughout the whole batch. Because of the excellent compatibility of Sunett with sugar alcohols, fruit jams and marmalades using these ingredients offer an outstanding taste.

Fruit preserves. Sunett can be used for the production of sugar-free or sugar-reduced fruit preserves. At the normal pH values for fruit preserves Sunett can be added even before pasteurization, as the sweetness is not impaired under the usual thermal treatment conditions. Sunett also withstands the usual storage periods without loss of sweetness. NutraSweet (r) is the commercial name of aspartame (APM), a new sweetener from G.D. SEARLE & Co. which can be used in the most foods in order to give the same taste as sugar. NutraSweet is about 180 to 200 times sweeter than sucrose (sugar) and this value depends on pH, temperature and the type of flavour.

NutraSweet can be used as mentioned above in a mix with Sunett or alone in all sugar-free or calorie-reduced fruit jams, marmalades and preserves. Like other sweeteners, NutraSweet does not promote tooth decay. International regulatory status. Both Sunett and NutraSweet are widely accepted by food laws in the majority of countries. The following are the main fruit products where Sunett is an accepted sweetener:

- low-joule prepared jelly: max. 500 mg/kg;
- canned fruit without added sugar: max. 500 mg/kg;
- beverages including calorie-reduced fruit nectars: max. 600 mg/l;
- calorie-reduced jams;
- calorie-reduced and dietetic fruit compotes;
- canned fruit and vegetables, fruit puree, jams and marmalades;
- juices, nectars and juice based beverages;
- jams, marmalades and related products: max. 300 mg/kg.

Packaging Materials

Requirements and Functions of Food Containers

The following are among the more important general requirements and functions of food packaging materials/containers:

a. they must be non-toxic and compatible with the specific foods;
b. sanitary protection;
c. moisture and fat protection;
d. gas and odour protection;
e. light protection;
f. resistance to impact;
g. transparency;
h. tamperproofness;
i. ease of opening;
j. pouring features;
k. reseal features;
l. ease of disposal;
m. size, shape, weight limitations;
n. appearance, printability;
o. low cost;
p. special features.

Primary and Secondary Containers

The terms primary and secondary containers have been used. Some foods are provided with efficient primary containers by nature, such as nuts, oranges, eggs and the like. In packaging these, we generally need only a secondary outer box, wrap, or drum to hold units together and give gross protection.

Other foods such as milk, dried eggs and fruit concentrates often will be filled into primary containers such as plastic liners which are then packaged within protective cartons or drums. In this case the secondary container provided by the carton or drum greatly minimises the requirements that must be met by the primary container. Except in special instances, secondary containers are not designed to be highly impervious to water vapour and other gases, especially at zones of sealing, dependence for this being placed upon the primary container.

Since primary containers by definition are those which come in direct contact with the food, we will be far more concerned with them than with secondary containers.

Hermetic Closure

Two conditions of the greatest significance in packaging are hermetic and non-hermetic closure. The term hermetic means a container which is absolutely impermeable to gases and vapours throughout its entirety, including its seams. Such a container, as long as it remains intact, will automatically be impervious to bacteria, yeasts, moulds, and dirt from dust and other sources since all of these agents are considerably larger than gas or water vapour molecules.

On the other hand, a container which prevents entry of micro-organisms, in many instances will be non-hermetic. A container that is hermetic not only will protect the product from moisture gain or loss, and from oxygen pick-up from the atmosphere, but is essential for strict vacuum and pressure packaging.

The most common hermetic containers are rigid metal cans and glass bottles, although faulty closures can make them non-hermetic. With very rare exceptions flexible packages are not truly hermetic for one or more of the following reasons.

First, the thin flexible films, even when they do not contain minute pinholes, generally are not completely gas and water-vapour impermeable although the rates of gas and water vapour transfer may be exceptionally slow; second, the seals are generally good but imperfect; and third, even where film materials may be gas- and water-vapour-tight, such as certain gages of aluminium foil, flexing of packages and pouches leads to minute pinholes and crease holes. Hermetic rigid aluminium containers can be readily formed without side seams or bottom end seams. The only seam then to make hermetic is the top end double seam, which may be closed on regular tin can sealing equipment.

Glass containers are hermetic provided the lids are tight. Lids will have inside rings of plastic or cork. Many glass containers are vacuum packed and the tightness of the cover will be augmented by the differential of atmospheric pressure pushing down the cover.

Crimping of the covers, as in the case of pop bottle caps which operate against positive internal pressure, also can make a gas-tight hermetic seal. But bottles fail more often than cans in becoming non-hermetic.

Protection of Food by Packaging Materials

Films and Foils; Plastics

Films and foils have different values for moisture and gas permeability, strength, elasticity, inflammability and resistance to insect penetration and many of these characteristics depend upon the film's thickness.

Important characteristics of the types of films and foils commonly used in food packaging are given in Table below. For the most part such films are used in the construction of inner containers. Since they are non-rigid, their main functions are to contain the product and protect it from contact with air or water vapour. Their capacity to protect against mechanical damage is limited, particularly when thin films are considered.

These materials can exist in many forms, depending upon such variables as identity and mixture of polymers, degree of polymerisation and molecular weight, spatial polymer orientation, use of plasticisers (softeners) and other chemicals, methods of forming such as casting, extrusion or calendering, etc.

One of the newer classes of plastic materials referred to as copolymers illustrate what can be done with mixtures of the basic units from which plastics are built. The term copolymer refers to a mixture of chemical species in the resin from which films and other forms can be made. The many variations possible make copolymers an important class of plastics to extend the range of useful food packaging applications.

Table: *Properties of packaging films*

Material	***Properties***
Paper	Strength; rigidity; opacity; printability.
Aluminium foil	Negligible permeability to water-vapour, gases and odours; grease proof, opacity and brilliant appearance; dimensional stability; dead folding characteristics.
Cellulose film (coated)	Strength; attractive appearance; low permeability to water vapour (depending on the type of coating used), gases, odours and greases; printability.
Polythene	Durability; heat-sealability; low permeability to water-vapour; good chemical resistance; good low-temperature performance.
Rubber hydrochloride	Heat-sealability; low permeability to water vapour, gases, odours and greases; chemical resistance.
Cellulose acetate	Strength; rigidity; glossy appearance; printability; dimensional stability.
Vinylidene chloride	Low permeability to water vapour, gases, copolymer odours and greases; chemical resistance; heat-sealability.
Polyvinyl chloride	Resistance to chemicals, oils and greases; heat-sealability.
Polyethylene terephtalate	Strength; durability; dimensional stability; low permeability to gases, odours and greases.

Source: FAD/WFP, (1970)

Plastic Sheets

- Cellophane paper can be used for packing of dried products, mainly for dried fruit leathers.
- Polyethylene sheets have a variety of uses. They are flexible, transparent and have a perfect resistance to low temperatures and impermeability to water vapour.

  An important advantage is that these sheets can be easily heat-sealed. Utilisation is in forms of sheets and bags. It is a good packing material for primary protection of dehydrated products. If a good protection is needed to prevent flavour and gas losses, it will be necessary to combine polyethylene with other materials.

Receptacles and Packagings in Plastic Materials

In this class there are three categories:

a. receptacles that can be heat treated: boxes, bottles and bags. Sterilisable bags used up to 120° C can be manufactured from same raw materials as described under plastic sheets and up to 100° C from cellophane. Polyethylene bags could be used to some extent for packing and pasteurization of sauerkraut.

b. receptacles that are not heat treated during processing of fruit and vegetables, also divided in bags and boxes. Bags are the most used type of packing from plastic materials and they are manufactured from polyethylene or cellophane; an important utilisation is for dried/dehydrated fruits and vegetables.

c. special packagings - which can be contacted (Criovac type) by action of heat once the finished product is already inside the pack and the air is evacuated.

Laminates

Various flexible materials such as papers, plastic films, and thin metal foils have different properties with respect to water vapour transmission, oxygen permeability, light transmission, burst strength, pin holes and crease hole sensitivity, etc. and so multi-layers or laminates of these materials which combine the best features of each are used. Commercial laminates containing up to as many as eight different layers are commonly custom designed for a particular product.

Laminations of different materials may be formed by various processes including bonding with a wet adhesive, dry bonding of layers with a thermoplastic adhesive, hot melt laminating where one or both layers exhibit thermoplastic properties, and special extrusion techniques. Such structured plastic films may be complete in themselves or be further bonded to papers or metal foils to produce more complex laminates.

Glass Containers

As far as food packaging is concerned, glass is chemically inert, although the usual problems of corrosion and reactivity of metal closures will of course apply. The principal limitation of glass is its susceptibility to breakage, which may be from internal pressure, impact, or thermal shock, all of which can be greatly minimised by proper matching of the container to its intended use and intelligent handling practices. Here consultation with the manufacturer cannot be over-stressed.

The heavier a jar or bottle for a given volume capacity the less likely it is to break from internal pressure. The heavier jar, however, is more susceptible to both thermal shock and impact breakage. Greater thermal shock breakage of the heavier jar is due to wider temperature differences which cause uneven stress between the outer and inner surfaces of the thicker glass. Greater impact breakage susceptibility of the heavier jar is due to the lower resilience of its thicker wall. Coatings of various types can markedly reduce each of these types of breakage. These coatings, commonly of special waxes and silicones, lubricate the outside of the glass. As a result, impact breakage is lessened by bottles and jars glancing off one another rather than sustaining direct hits when they are in contact in high speed filling lines.

Surface coating after annealing protects glass surfaces from many of the minute scratches appearing in normal handling after annealing ovens; surface coating also improves the high gloss appearance of glass containers and is said to decrease the noise from glass to glass contact at filling lines. With regard to thermal shock, it is good practice to minimise temperature differences between the inside and outside of glass containers whenever possible. Some manufacturers will recommend that a temperature difference of 44° C (80° F) between the inside and the outside not be exceeded. This requires slow warming of bottles before use for a hot fill and partial cooling before such containers are placed under refrigeration.

Classification

Glass used for receptacles in fruit and vegetable processing is a carefully controlled mixture of sand, soda ash, limestone and other materials made molten by heating to about 1500° C (2800° F).

Main classes of glass receptacles are:

a. jars which are resistant to heat treatments,

b. jars, glasses, etc. for products not submitted to heat treatment (marmalades, acidified vegetables, etc.);

c. glass bottles for pasteurized products (tomato juice, fruit juices, etc.) or not pasteurized (syrups) and

d. receptacles with higher capacity (flasks, etc.)

Jars for Sterilised/Pasteurized Canned Products

These receptacles may replace metal cans, taking into considering both the advantages and disadvantages they present. Advantages are:

they do not react to food content; they are transparent and can be manufactured in various shapes; they use cheap raw materials and are reusable. Disadvantages: heavier than metal can of same capacity; fragile; lower thermal conductance and a limited resistance to thermal shocks.

Receptacles in this category must assure a perfect hermeticity after their pasteurization/sterilisation and cooling and this has to be achieved by the use of metallic (or glass) caps and specific materials for tightness. Taking into account the receptacles' closure method, there are two categories of receptacles:

a. glass jars with mechanical closure;

b. glass jars with pneumatic closure.

Jars for Products without Heat Treatment

For marmalades, jellies and jams glass jars with non hermetic closures using metal, glass or rigid plastic caps are used; however for these products the receptacles mentioned above may also be used. The use of jars with pneumatic closure presents the advantage that some products (e.g. marmalades, jams) can be filled hot and therefore sterile in receptacles. Pneumatic closing generally protects against negative air action which is in this case eliminated from receptacles.

Glass Bottles

These receptacles are widely used both for

a) finished products which need pasteurization (ea. tomato juice, Knit juices, etc.) and for

b) those which are preserved as such (ea. fruit syrups).

Glass bottles in category a) are closed hermetically with metallic caps, provided with special materials for tightness. For glass bottles in category b) various corks, and aluminium caps with tightness materials may be used.

Glass Receptacles with High Capacity

In this category there are glass flasks with 3 and 10 litre capacity which can be hermetically closed by a SKO caps system and are resistant to product pasteurization (ea. tomato juice).

As bigger receptacles it is possible to use glass demijohns with usual capacity of 25 and 50 litre; these receptacles are used for preservation of fruit juices by warm process. Closing is performed with flexible rubber hoods.

Paper Packaging

As primary containers few paper products are not treated, coated or laminated to improve their protective properties. Paper from wood pulp and reprocessed waste paper will be bleached and coated or impregnated with such materials as waxes, resins, lacquers, plastics, and laminations of aluminium to improve water vapour and gas impermeability, flexibility, tear resistance, burst strength, wet strength, grease resistance, sealability, appearance, printability, etc.

Paper Sheets

- Kraft paper is the brown unbleached heavy duty paper commonly used for bags and for wrapping; it is seldom used as a primary container;
- parchment paper: acid treatment of paper pulp modifies the cellulose and gives water and oil resistance and considerable wet strength to this type of packaging material;
- glassine-type papers are characterised by long wood pulp fibres which impart increased physical strength;
- paper with plastic material sheets.

Receptacles from Paper or Cardboard

(paper = 8 to 150 g/m^2; cardboard = 150 to 450 g/m^2).

"Tin can"/tinplate

The "tin can" is a container made of tinplate.

Tinplate, a rigid and impervious material, consists of a thin sheet of low carbon steel coated on both sides with a very thin layer of tin. It can be produced by dipping sheets of mild steel in molten tin (hot-dipped tinplate) or by the electro-deposition of tin on the steel sheet (electrolytic tinplate). With the latter process it is possible to produce tinplate with a heavier coating of tin on one surface than the other (differentially coated). Tin is not completely resistant to corrosion but its rate of reaction with many food materials is considerably slower than that of steel. The effectiveness of a tin coating depends on:

a. its thickness which may vary from about 0.5 to 2.0 µm (20 to 80 x 10(-6) in.);
b. the uniformity of this thickness;
c. the method of applying the tin which today primarily involves electrolytic plating;
d. the composition of the underlying steel base plate;

e. the type of food, and

f. other factors.

Some canned vegetables including tomato products actually owe their characteristics flavours to a small amount of dissolved tin, without which these products would have an unfamiliar taste. On the other hand, where tin reacts unfavourably with a particular food the tin itself may be lacquer coated.

The thickness of tinplate sheets may vary from 0.14 mm to 0.49 mm and is determined by weighing a sheet of known area and calculating the average thickness. Tinplate sheets may be lacquered after fabrication to provide an internal or external coating to protect the metal surface from corrosion by the atmosphere or through reaction with the can contents. They may also be printed by lithography to provide suitable instructions or information on containers fabricated from tinplate sheets (otherwise paper labels can be attached to the outer tinplate surface).

Under normal conditions the presence of the tin coating provides a considerable degree of electrochemical protection against corrosion, despite the fact that in both types of tinplate the tin coating is discontinuous and minute areas of steel base plate are exposed. With prolonged exposure to humid conditions, however, corrosion may become a serious problem. The coatings not only protect the metal from corrosion by food constituents but also protect the foods from metal contamination, which can produce a host colour and flavour reactions depending upon the specific food. Particularly common are dark coloured sulphides of iron and tin produced in low acid foods that liberate sulphur compounds when heat processed, and bleaching of red plant pigments in contact with unprotected steel, tin, and aluminium.

Table: *General types of can coatings*

Coating	***Typical uses***	***Type***
Fruit enamel	Dark coloured berries, cherries and other fruits requiring protection from metallic salts	Oleoresinous
C-enamel	Corn, peas and other sulphur-bearing products	Oleoresinous w. suspended zinc oxide
Citrus enamel	Citrus products and concentrates oleoresinous	Modified

Contd...

Coating	***Typical uses***	***Type***
Beverage can enamel	Vegetable juices; red fruit juices; highly corrosive fruits; non-carbonated beverages	Two-coated w. resinous base coat and vinyl top coat

Fruit Specific Preservation Technologies

Fruit Quality

Fruit quality goes back to tree stock, growing practices and weather conditions. Closer to the shipper and processor, however, are the degrees of maturity and ripeness when picked and the method of picking or harvesting. There is a distinction between maturity and ripeness of a fruit. Maturity is the condition when the fruit is ready to eat or if picked will become ready to eat after further ripening. Ripeness is that optimum condition when colour, flavour and texture have developed to their peak.

Some fruit is picked when it are mature but not yet ripe. This is especially true of very soft fruit like cherries and peaches, which when fully ripe are so soft as to be damaged by the act of picking itself. Further, since many types of fruit continue to ripen off the tree, unless they were to be processed quickly, some would become overripe before they could be utilised if picked at peak ripeness.

From a technological point of view, fruit characterisation by species and varieties is performed on the basis of physical as well chemical properties: shape, size, texture, flavour, colour/pigmentation, dry matter content (soluble solids content), pectic substances, acidity, vitamins, etc. These properties are directly correlated with fruit utilisation. Table below shows which of the above mentioned properties have a major impact on the finished products obtained by fruit processing.

Table: *Optimal use of fruits as a function of their properties*

Processed	***Organoleptic (Sensory) Properties***				***Chemical Composition***		
Finished Products	***Shape***	***Texture***	***Flavour***	***Taste***	***Acidity***	***Sugars***	***Pectic Sust.***
Dried Fruits	+ +	+ +		+ +		+ +	
Fruit Juices			+ +	+ +	+ +		
Marmalade			+ +	+ +			+ +

Contd...

Processed	*Organoleptic (Sensory) Properties*				*Chemical Composition*		
Finished Products	***Shape***	***Texture***	***Flavour***	***Taste***	***Acidity***	***Sugars***	***Pectic Sust.***
Jams	+ +	+ +	+ +	+ +			
Jellies	+ +	+ +	+ +	+ +			+ +
Fruit Paste				+ +		+ +	

When to Pick

The proper time to pick fruit depends upon several factors; these include variety, location, weather, ease of removal from the tree (which change with time), and purpose to which the fruit will be put.

For example, oranges change with respect to both sugar and acid as they ripen on the tree; sugar increases and acid decreases. The ratio of sugar to acid determines the taste and acceptability of the fruit and the juice. For this reasons, in some countries there are laws that prohibit picking until a certain sugar-acid ratio has been reached. In the case of much fruit to be canned, on the other hand, fruit is picked before it is fully ripe for eating since canning will further soften the fruit.

Quality Measurements

Many quality measurements can be made before a fruit crop is picked in order to determine if proper maturity or degree of ripeness has developed. Colour may be measured with instruments or by comparing the colour of fruit on the tree with standard picture charts.

Texture may be measured by compression by hand or by simple type of plungers.

As fruit mature on the tree its concentration of juice solids, which are mostly sugars, changes. The concentration of soluble solids in the juice can be estimated with a refractometer or a hydrometer. The refractometer measures the ability of a solution to bend or refract a light beam which is proportional to the solution's concentration. A hydrometer is a weighted spindle with a graduated neck which floats in the juice at a height related to the juice density.

The acid content of fruit changes with maturity and affects flavour. Acid concentration can be measured by a simple chemical titration on the fruit juice. But for many fruits the tartness and flavour are really affected by the ratio of sugar to acid.

Percentage of soluble solids, which are largely sugars, is generally expressed in degrees Brix, which relates specific gravity of a solution to an equivalent concentration of pure sucrose. In describing the taste of tartness of several fruits and fruit juices, the term "sugar to acid ratio" or "Brix to acid ratio" are commonly used. The higher the Brix the greater the sugar concentration in the juice; the higher the "Brix to acid ratio" the sweeter and lees tart is the juice.

Harvesting and Preprocessing

The above and other measurements, plus experience, indicate when fruit is ready for harvesting and subsequent processing. A large amount of the harvesting of most fruit crops is still done by hand; this labour may represent about half of the cost of growing the fruit.

Therefore, mechanical harvesting is currently one of the most active fields of research for the agricultural engineer, but also requires geneticists to breed fruit of nearly equal size, that matures uniformly and that is resistant to mechanical damage.

A correct manual harvesting includes some simple but essential rules:

- the fruit should be picked by hand and placed carefully in the harvesting basket; all future handling has to be performed carefully in order to avoid any mechanical damage;
- the harvesting basket and the hands of the harvester should be clean;
- the fruit should be picked when it is ready to be able to be processed into a quality product depending on the treatment which it will undergo.

It is worth emphasising the fact that the proximity of the processing centre to the source of supply for fresh raw materials presents major advantages; some are as follows:

- possibility to pick at the best suitable moment;
- reduction of losses by handling/transportation;
- minimises raw material transport costs;
- possibility to use simpler/cheaper receptacles for raw material transport.

Once it has left the tree, the organoleptic properties, nutritional value, safety and aesthetic appeal of the fruit deteriorates in varying degrees. The major causes of deterioration include the following:

a. growth and activity of micro-organisms;
b. activities of the natural food enzymes;
c. insects, parasites and rodents;
d. temperature, both heat and cold;
e. moisture and dryness;
f. air and in particular oxygen;
g. light and
h. time.

The rapidity with which foods spoil if proper measures are not taken is indicated in Table below.

Table: *Useful storage life of some food products*

Food Products	***Generalized Storage Life (Days) at 21°C (70°F)***
Animal Flesh, Fish, Poultry	1-2
Dried, salted, smoked meat and fish	360 and more
Fruits	1-7
Dried Fruits	360 and more
Leafy Vegetables	1-2
Root Crops	7-20

Source: Desrosier and Desrosier (1977)

Reception - Quality and Quantity

Fruit reception at the processing centre is performed mainly for following purposes:

- checking of sanitary and freshness status;
- control of varieties and fruit wholeness;
- evaluation maturity degree;
- collection of data about quantities received in connection to the source of supply: outside growers/farmers, own farm.

Variety control is needed in order to identify that the fruit belongs to an accepted variety as not all are suitable for different technological processes. Fruit maturity degree is significant as industrial maturity is required for some processing/preservation methods while for others there is the need for an edible maturity when the fruit has full taste and flavour.

Special attention is given to size, appearance and uniformity of fruit to be processed, mainly in the form of fruit preserved with sugar using whole/half fruits ("with fruit pieces").

Some laboratory control is also needed, even if it not easy to precisely establish the technological qualities of fruit because of the absence of enough reliable rapid analytical methods able to show eventual deterioration. The only reliable method for evaluating the quality is the combination of data obtained through organoleptic/taste controls and by simple analytical checks which are possible to perform in a small laboratory: percentage of soluble solids by refractometer, consistency/texture measured with simple penetrometers, etc.

Temporary Storage before Processing

This step has to be as short as possible in order to avoid flavour losses, texture modification, weight losses and other deterioration that can take place over this period.

Some basic rules for this step are as follows:

- keep products in the shade, without any possible direct contact with sunlight;
- avoid dust as much as possible;
- avoid excessive heat;
- avoid any possible contamination;
- store in a place protected from possible attack by rodents, insects, etc.

Cold storage is always highly preferred to ambient temperature. For this reason a very good manufacturing practice is to use a cool room for each processing centre; this is very useful for small and medium processing units as well.

Washing

Harvested fruit is washed to remove soil, micro-organisms and pesticide residues. Fruit washing is a mandatory processing step; it would be wise to eliminate spoiled fruit before washing in order to avoid the pollution of washing tools and/or equipment and the contamination of fruit during washing.

Washing efficiency can me gauged by the total number of micro-organisms present on fruit surface before and after washing - best result are when there is a six fold reduction. The water from the final wash should be free from moulds and yeast; a small quantity of bacteria is acceptable.

Fruit washing can be carried out by immersion, by spray/showers or by combination of these two processes which is generally the best solution: pre-washing and washing.

Some usual practices in fruit washing are:

- addition of detergents or 1.5% HCl solution in washing water to remove traces of insect-fungicides;
- use of warm water (about 50°C) in the pre-washing phase;
- higher water pressure in spray/shower washers.

Washing must be done before the fruit is cut in order to avoid losing high nutritive value soluble substances (vitamins, minerals, sugars, etc.).

Sorting

Fruit sorting covers two main separate processing operations:

a. removal of damaged fruit and any foreign bodies (which might have been left behind after washing);
b. qualitative sorting based on organoleptic criteria and maturity stage.

Mechanical sorting for size is usually not done at the preliminary stage. The most important initial sorting is for variety and maturity. However, for some fruit and in special processing technologies it is advisable to proceed to a manual dimensional sorting (grading).

Trimming and Peeling (Skin Removal)

This processing step aims at removing the parts of the fruit which are either not edible or difficult to digest especially the skin. Up to now the industrial peeling of fruit and vegetables was performed by three procedures:

a. mechanically;
b. by using water steam;
c. chemically; this method consists in treating fruit and vegetables by dipping them in a caustic soda solution at a temperature of 90 to 100° C; the concentration of this solution as well as the dipping or immersion time varying according to each specific case.

Cutting

This step is performed according to the specific requirements of the fruit processing technology.

Heat Blanching

Fruit is not usually heat blanched because of the damage from the heat and the associated sogginess and juice loss after thawing.

Instead, chemicals are commonly used without heat to inactivate the oxidative enzymes or to act as antioxidants and they are combined with other treatments.

Ascorbic/Citric Acid Dip

Ascorbic acid or vitamin C minimises fruit oxidation primarily by acting as an antioxidant and itself becoming oxidised in preference to catechol-tannin compounds. Ascorbic acid is frequently used by being dissolved in water, sugar syrup or in citric acid solutions.

It has been found that increased acidity also helps retard oxidative colour changes and so ascorbic acid plus citric acid may be used together. Citric acid further reacts with (chelates) metal ions thus removing these catalysts of oxidation from the system.

Sulphur Dioxide Treatment

Sulphur dioxide may function in several ways:

- sulphur dioxide is an enzyme poison against common oxidising enzymes;
- it also has antioxidant properties; i.e., it is an oxygen acceptor (as is ascorbic acid);
- further SO2 minimises non enzymatic Maillard type browning by reacting with aldehyde groups of sugars so that they are no longer free to combine with amino acids;
- sulphur dioxide also interferes with microbial growth.

In many fruit processing pre-treatments two factors must be considered:

a. sulphur dioxide must be given time to penetrate the fruit tissues;

b. SO_2 must not be used in excess because it has a characteristic unpleasant taste and odour, and international food laws limit the SO_2 content of fruit products, especially of those which are consumer oriented (e.g. except semi-processed products oriented to further industrial utilisation).

Commonly a 0.25 % solution (except for semi-processed fruit products which are industry oriented and use a 6% solution) of SO_2 or its SO_2 equivalent in the form of solutions of sodium sulphite, sodium bisulphite or sodium/potassium metabisulphite are used. Fruit slices are dipped in the solution for about two to three minutes and then removed so as not to absorb too much SO2. Then the slices are allowed to stand for about one to two hours so that the SO2 may penetrate throughout the tissues before processing. Sulphur dioxide is also used in fruit juice production to minimise oxidative changes where relatively low heat treatment is employed so as not to damage delicate juice flavour. Dry sulphuring is the technological step where fruit is exposed to fumes of SO2 from burning sulphur or from compressed gas cylinders; this treatment could be used in the preparation of fruits (and some vegetables) prior to drying/dehydration.

Sugar Syrup

Sugar syrup addition is one of the oldest methods of minimising oxidation. It was used long before the causative reactions were understood and remains today a common practice for this purpose. Sugar syrup minimises oxidation by coating the fruit and thereby preventing contact with atmospheric oxygen. Sugar syrup also offers some protection against loss of volatile fruit esters and it contributes sweet taste to otherwise tart fruits. It is common today to dissolve ascorbic acid and citric acid in the sugar syrup for added effect or to include sugar syrup after an SO_2 treatment.

Fresh Fruit Storage

Some fruit species and specially apples and pears can be stored in fresh state during cold season in some countries' climatic conditions. Fruit for fresh storage have to be autumn or winter varieties and be harvested before they are fully mature. This fruit also has to be sound and without any bruising; control and sorting by quality are mandatory operations. Sorting has to be carried out according to size and weight and also by appearance; fruit which is not up to standard for storage will be used for semi-processed product manufacturing which will be submitted further to industrial processing.

Harvested fruit has to be transported as soon as possible to storage areas. Leaving fruit in bulk in order to generate transpiration is a bad practice as this reduces storage time and accelerates maturation processes during storage. In order to store large quantities of fruit, silos have to be built.

Technology of Semi-processed Fruit Products

The semi-processed fruit products are manufactured in order to be delivered to industry processing centres (in the fruit producing country itself or in importing countries) where they will be further manufactured in consumer oriented finished products: jams, jellies, syrups, fruits in syrup, etc.

In the practice of semi-processed fruit products and for the purpose of this document the following categories are defined:

a. fruit "pulps": semi-processed products, not refined, obtained by mechanical treatment (or, less often, by thermal treatment) of fruit followed by their preservation. Either whole fruit, halves or big pieces are used which enables easy identification of the species. "Pulps" can be classified in boiled or non boiled (raw).
b. fruit "purées-marks": semi-processed products obtained by thermal and mechanical treatment or, very rare, raw and then refined, operations by which all nonedible parts (cores, peels, etc.) are removed. "Purées-marks" are classified in boiled (the more usual case) and non boiled (raw).
c. semi-processed juices: products obtained by cold pressure or very rare by other treatments (diffusion, extraction, etc.) followed by the preservation.

Technical Processes for Preservation of Semi-processed Fruit Products

Preservation can be achieved by chemical means, by freezing or by pasteurization. The choice of preservation process for each individual case is a function of the semi-processed product type and the shelf life needed.

Chemical preservation. In many countries, in practice, this is carried out with sulphur dioxide, sodium benzoate, formic acid and, on a small scale, with sorbic acid and sorbates.

Preservation with sulphur dioxide is a widespread process because of its advantages: universal antiseptic action and very economic application. The drawbacks of SO_2 are: SO_2 turn firms the texture of some fruit species (pomaces), desulphiting is not always complete and recolouring of red fruits is not always complete after desulphitation.

Practical preservation dosage levels with SO_2 for about 12 months is 0.18-0.20% SO_2 (with respect to the product to be preserved).

This level could be reduced to 0.09% SO_2 for 3 months and to 0.12% SO_2 for 6 months preservation.

The preservation with sulphur dioxide is in use mainly for "pulps" and for "purées-marks".

Chemical preservation can be performed from a practical point of view by the utilisation of 6% SO2 water solutions or by direct introduction of sulphur dioxide gas in the product (for "purées-marks"). The preparation of 6% SO2 solutions is done by bubbling the gas from cylinders in cold water; from a 50 kg SO_2 compressed gas cylinder results 830 l of 6% SO2 solution.

These SO_2 solutions have to be stored in cool places, in closed receptacles and with periodic concentration control/check by titration or by density measurements approximate results. Preservation with sodium benzoate has the following advantages: it does not firm up the texture and does not modify fruit colour. The disadvantages are: it is not a universal antiseptic, its action needs an acid medium and the removal is partial. Sodium benzoate is in use for "pulps" and for "purées-marks" but less for fruit juices.

Practical dosage level for 12 months preservation is 0.18-0.20 % sodium benzoate, depending on the product to be preserved. Sodium benzoate is used as a solution in warm water; the dissolution water level has to be at maximum 10% reported to semi-processed product weight. Formic acid preservation is performed mainly for semi-processed fruit juices at a dosage level of 0.2 % pure formic acid (100%). Formic acid is an antiseptic effective against yeasts, does not influence colour of products and is easily removed by boiling.

Formic acid could be diluted with water in order to insure a homogeneous distribution in the product to be preserved; water has to be at maximum 5 % of the product weight. Because of a potential effect of pectic substance degradation, formic acid is less in use for "pulps" and "purées-marks" preservation. Sorbic acid used as potassium sorbate (easily water soluble) can be used for preservation of fruit semi-processed products at a dosage level of 0. 1% maximum. Advantages of sorbates are: they are completely harmless and without any influence on the organoleptic properties of semi-processed fruit products.

Preservation by pasteurization. As fruit has a low pH, preservation of semiprocessed fruit products could also be performed by pasteurization (heat treatment step at maximum temperature of 100° C), the length of this step varying with the size of the receptacles.

The advantages of this type of treatment are: hygienic process, which assure a long term preservation; the disadvantages are: need for air tight receptacles, and pectic substances could begin to deteriorate if the thermal treatment is too long. Thermal preservation of fruit semi-processed products could also be done by a "selfpasteurization": very hot semi-processed products are filled into receptacles (e.g. metal cans) which are sealed and then inverted in order to sterilise the air which goes through the hot fruit mass.

Preservation by freezing. This is done on an industrial scale in some countries and can be done with or without sugar addition. The advantages of this process are: absence of added substances; very good preservation of quality of fruit constituents (pectic substances, vitamins, etc.) and good preservation of organoleptic properties (flavour, taste, colour). Freezing is done at about -20 to -30° C and storage at -10 to -18° C.

Freezing is applied mainly to semi-processed fruit products aimed at very high quality and high cost finished products.

Technological flow-sheet for semi-processed fruit "pulps": chemical preservation.

Sorting is needed in order to remove sub-standard fruit (with moulds, with diseases, etc.) and all foreign bodies.

Washing is obligatory in order to remove all impurities which cannot be eliminated at the processing step in finished products.

Coring and Cutting, mainly for pomace fruits, has as main objective a better utilisation of preservation "space" in receptacles and is not mandatory; this will be defined by customer/supplier agreements/ standards. This operation is preferably performed by mechanical means.

Preservation is carried out with the 6% SO2 solution which is added to the prepared fruits (placed in bulk in receptacles) in the quantity needed to obtain the preservation dosage level. For a better/ homogeneous preservative distribution, the initial 6% SO2 solution could be diluted with water; however, the diluted solution (which will be filled in receptacles) has to be at a dosage level of less than 10% of the semi-processed product weight.

For some soft fruit, especially strawberries, preservation is done with a mix of 6% SO2 solution and calcium bisulphite solution (containing also 6% SO2).

Preparation of calcium bisulphite solution is done by the introduction of 30 kg of CaO in 1 m^3 SO2 solution and mixing up to clarification. The resulting solution is mixed with the initial 6% SO2 solution, generally in a 1:1 ratio, but the ratio can be adapted to the fresh fruit texture. Firming of soft fruit texture by this treatment is based on the formation of calcium pectate with pectic substances from fruit tissues.

In the case of sodium benzoate, formic acid or potassium sorbate, the dosage levels to be used are as indicated above with the rule that it is not allowed to add more that 10% liquid in receptacles on the prepared fruits. Preservation by pasteurization or "self-pasteurization" will need as additional steps: a) boiling with a minimum water addition (maximum 10%); b) filling of receptacles; c) hermetic closing followed by d) pasteurization or "self-pasteurization".

Some general technical data for the preparation of chemical preserved semi-processed fruit "pulps" are seen in Table below.

Table: *General technical processing data for semi-processed fruit "pulps"*

Fruit species	Preliminary operations	Preservation means
Apples, pears, quinces	Sorting, washing, coring, cutting	Sulphur dioxide
Prunes, peaches, wax cherries, apricots	Sorting, washing, stone removal (pitting)	Sulphur dioxide or sodium benzoate
Cherries	Sorting, washing	Sulphur dioxide or sodium benzoate, sometimes with calcium bisulphite addition
Strawberries	Sorting, washing	Sulphur dioxide in mix with calcium bisulphite
Wild berries	Sorting, washing	Sulphur dioxide or sodium benzoate; in some cases with calcium bisulphite addition

Technological Flow-sheet for Semi-processed "Purée-marks"

The general technological flow-sheet includes the following operations:

- Sorting and Washing are obligatory and are carried out in a similar manner as for "pulps".
- Heat Treatment/Boiling is needed in order to soften the fruit tissues before refining. For some fruits as strawberry and wild berries, this step is not done and fruits are refined "raw" in order to preserve their flavour.
- Pulping is performed with specific equipment - refiners or pulpers - which eliminate seeds, pits and other non edible parts (peels, cores, etc.).
- Preservation is carried out by chemical means, by freezing or by pasteurization.

The general technical/processing data for the manufacture of "purée-marks" are seen in Table below.

Table: *General technical processing data for semi-processed "purées - marks"*

Fruit species	Preliminary operations	Preservation means
Apples, pears, quinces	Sorting, washing, boiling, refining	Sulphur dioxide; in less frequent cases formic acid or sodium benzoate
Prunes, peaches, wax cherries, apricots,	Sorting, washing, boiling, refining	Sulphur dioxide, formic acid or sodium benzoate; cherries freezing with or without sugar; self-pasteurization
Strawberry, wild berries	Sorting, washing, refining	Chemical preservation, freezing or self pasteurization

Fruit Sugar Preserves Technology; Jams, Jellies, Marmalade, Fruit Paste

As a overall rule of thumb, a sugar concentration of about 60% in finished or processed fruit products generally insures their preservation. Preservation is not only determined by the osmotic pressure of sugar solutions but also by the water activity values in the liquid phase, which can be lowered by sugar addition; and by evaporation down to 0.848 aw; this value however does not protect products from mould and osmophile yeast attack.

Maximum saccharose concentration that can be achieved in the liquid phase of the product is 67.89%; however higher total sugar quantities (up to 70-72%) found in products are explained by an increased reducing sugar solubility resulting from saccharose inversion.

Jams

The preservation of fruit by jam making is a familiar process carried out on a small scale by housewives in many parts of the world. Factory jam making has become a highly complex operation, where strict quality control procedures are employed to ensure a uniform product, but the manufacturing operations employed are in essence the same as those employed in the house.

Fresh or pre-cooked fruit is boiled with a solution of cane or beet sugar until sufficient water has been evaporated to give a mixture which will set to a gel on cooling and which contains 32-34% water.

Gel formation is dependent on the presence in the fruit of the carbohydrate pectin, which at a pH of 3.2 - 3.4 and in the presence of a high concentration of sugar, has the property of forming a viscous semi-solid. During jam boiling, all micro-organisms are destroyed within the product, and if it is filled hot into clean receptacles which are subsequently sealed, and then inverted so that the hot jam contacts the lid surface, spoilage by micro-organisms will not take place during storage. The composition of jam made from stone fruit and berry fruit is shown in Table below. About 30% of the vitamin C present in fresh fruit is destroyed during the jam-making process, but that which remains in the finished product is stable during storage.

The high moisture content of jam (equivalent to an equilibrium relative humidity of about 82%) makes it susceptible to mould damage once the receptacle has been opened and exposed from some time to the air. No problems of microbiological spoilage are likely to arise in the canned product during storage.

Table: *Composition of some fruit jams*

Type of Jam	Moisture content %	Sugar (as invert sugar, %)	Vitamin C mg/100 g
Jam made from berry fruits: strawberry, raspberry, etc.	29.8	69.0	10 - 25
Jam made from stone fruits: apricot, peach, etc.	29.6	69.3	10 - 35

Source: FAD/WFP, 1970

Marmalade

This sugar preserve is defined as "semisolid or gel-like product prepared from fruit ingredients together with one or more sweetening ingredients and may contains suitable food acids and food pectins; the ingredients are concentrated by cooking to such a point that the TSS - Total Soluble Solids - of the finished marmalade is not below 65%".

Fruit Paste

Fruit paste is a product obtained in the same way as special non-gelified fruit marmalade but with a lower water content - about 25% TSS in fruit paste.

Lowering water content could be achieved by continuing boiling of the product or by drying the product by natural or artificial drying. An example of paste without sugar is the sun dried apricot or prune paste. General procedure for the preparation of jams, jellies and marmalade:

a. Boil the pulp or the juice (with water when necessary)
b. Add the pectin
   - to the batch while stirring very vigorously
   - Pectin which has previously been mixed with 5 times its weight in sugar taken from the recipe)
c. Boil for about 2 minutes to assure a complete dissolution
d. Add the sugar while keeping the batch boiling
e. Boil down quickly to desired Brix
f. Add the acid (usually citric acid) and remove the froth
g. Fill hot into the (previously cleaned) jars and close
h. Invert the jars for three minutes to pasteurize the cover.

* Note: the pectin in solution can also be added at the end of the step (e) and has to be prepared as follows: use a strong blender. For one litre of water add slowly into the blender 25g of pectin mixed with 100 g of sugar taken from the recipe.

Basic recipes

The following recipes must be considered only as guidelines because the composition of the fruit can vary; also the taste of the consumers varies concerning the consistency, the sweetness and acidity. Before starting to make jam it is important to know the yield to settle the question on containers. The calculation is made as follows:

$$\frac{\text{Total Soluble Solids content of all the raw ingredients}}{\text{Percentage of Total Soluble Solids in the Final Product}} \times 100$$

In these basic recipes it is assumed that the fruits are poor in pectin content.

Recipe 1. Fruit: sugar = 50:50; desired Brix in the finished product is 68.

Soluble Solids:

10 kg of fruit at 10% TSS	1.000 kg
10 kg of sugar	10.000 kg
60 g of pectin (grade 200)	0.060 kg
55 g of citric acid	0.055 kg
	11.115 kg

$$\text{Yield} = \frac{11.115 \times 100}{68} = 16.4\,\text{kg}$$

Recipe 2. Fruit: sugar = 45:55; desired Brix in the finished product is 68.

Soluble Solids:

d10 kg of fruit at 10% TSS	1.000 kg
2.5 litre of water	-
12.2 kg of sugar	12.200 kg
65 g of pectin grade 200	0.060 kg
60 g of citric acid	0.060 kg
	13.325 kg

$$\text{Yield} = \frac{13.325 \times 100}{68} = 19.6\,\text{kg}$$

Recipe 3. Fruit: sugar = 40:60; desired Brix in the finished product is 68.

Soluble Solids:

d10 kg of fruit at 10% TSS	1.000 kg
3.3 litre of water	-
15 kg of sugar	15.000 kg
85 g of pectin grade 200	0.085 kg
80 g of citric acid	0.080 kg
	16.165 kg

$$\text{Yield} = \frac{16.165 \times 100}{68} = 23.8\,\text{kg}$$

Various factors must be taken into account:

1. Size of the container: the quantity of pectin indicated in the recipes is valid for containers of 1 kg or less.

If container capacity is between	*Increase pectin by*
1 kg and 2.5 kg	5%
2.5 kg and 5.0 kg	10%
5.0 kg and 10.0 kg	20%
10.0 kg and 20.0 kg	30%

2. Finishing point: the quantity of pectin is given for a final Brix - Total Soluble Solids (TSS) of 68%.

   If the final Brix is 66 increase the pectin by 5%

   " " is 65 " " " by 10%

   is 64 " " " by 15%

   " " is 62 " " " by 20%

   " " is 60 " " " by 30%

3. Acidic taste. If the product is too acid, replace the citric acid by tartaric acid (63% of the amount of citric acid).
4. Formation of clots: If batch clots, it is probably due to the pH being too low or the or TSS being too high; correct accordingly.
5. Formation of liquid at the surface: if liquid forms on the surface, it is probably due to too low a pH or too low pectin content.
6. Crystallisation: a) if liquid forms on the surfaces, then the pH is too low; reduce the acid content; b) if liquid does not form on the surface, then TSS or pH is too high.
7. Formation of mould: the TSS is probably below 68 deg. Brix. The filling may have been done at a low temperature. If the containers are large, wait until they are cold before closing.
8. Wrong batch: dilute the jam with water to 30% TSS; cook briefly. Add this diluted jam to a new batch but in a ratio not exceeding 10%.

Processing of Pineapple-papaya Jam

The fruit should be prepared as per previous instructions. For pineapples, the ends are removed and discarded; the cores and outer parts of the fruits are also removed. The fruit cylinders obtained are pulped through a special extractor (Fitzpatrick communiting machine) equipped with a 0.40-in screen sieve; the pulp thus obtained is used for making jam.

The papaya are prepared by hand-peeling the fruit; the fruit is then halved and the seeds removed. It is then pulped in the communiting machine using a 0.40-in screen sieve. When ginger root is used as flavouring, it is peeled and macerated in a Kenwood blender to a very fine consistency.

A typical formula for a pineapple-papaya jam (50:50 ratio) with ginger flavouring is given as follows:

Pineapple pulp	25.0
Papaya pulp	25.0 pounds
Cane sugar	50.0
Apple pectin (150 grade)	6.0 ounces
Citric acid	6.4
Fresh ground ginger	7.5

Processing is carried out in the following way:

The weighed fruit pulp is placed in a stainless steel steam-jacketed kettle and heated to about 110°F under constant stirring.

When the product reaches this temperature, the heat is turned off. The pectin (mixed in about ten times its weight with some of the weighed sugar), is then mixed into the fruit pulp, stirring constantly in order to prevent the pectin from clotting.

When the pectin has dissolved, the remainder of the sugar is added and dissolved completely in the mixture. The heat is then turned on and the jam mixture is stirred constantly until it starts boiling vigorously. During the remainder of the cooking, the product is stirred occasionally. Near the finishing point (approximately 221° F), the citric acid and the ginger (if it is used) are also added.

Determination of the finishing point is done by removing samples at intervals, cooling, and reading the soluble solids by means of a refractometer equipped with a Brix scale. After the jam reaches the proper Total Soluble Solids content, the heat is turned off and the surface scum/foam is removed.

The jam then is quickly put into receptacles which have been cleaned and sterilised with boiling in water for 30 minutes. The filling operation is done rapidly in order to prevent the temperature of the jam from falling below 190° F.

After filling, sterilised lids (boiled for 30 minutes in water) are placed on the receptacles and they are then sealed.

After this operation the receptacles are inverted for about 3 minutes to insure that the lids are sterilised. The receptacles are then placed upright.

At this stage it is not necessary to do any further processing, therefore the receptacles are cooled in running cold water until they reach a temperature slightly above room temperature. They are then dried in air and labelled.

Evaluation of Finished Products

During production at medium/large scale, it is recommended that quality controls be performed during manufacturing. After ten weeks of storage at room temperature it is recommended that an examination of finished products be performed.

The receptacles are opened and contents carefully emptied on to enamel trays without disturbing the formation of the jam.

The empty cans (if metal cans were used) are then inspected for signs of corrosion. Factors other than flavour include colour, appearance, syrup separation, firmness and spreading quality.

For flavour, jam is tested on pieces of bread. Samples are taken for measurement of pH (with a glass electrode pH meter) and Total Soluble Solids (with a refractometer equipped with a Brix scale).

This evaluation enables to have a quality check during product shelf life and to obtain data needed for necessary improvements of future productions.

For pineapple-papaya jam, products made with 30% pineapple and 70% papaya with added ginger has the highest score for flavour. The use of plain tin cans causes corrosion problems which is not the case when acid resistant lacquer cans are used.

Pineapple Jam Making

1. Boil 40 lb. of pulp and 12 lb. of water.
2. Add 225 g of pectin to the batch while stirring rapidly.
3. Boil for about 90 sec to assure complete dissolution.
4. Add 60 lb. of sugar gradually if possible in several portions, while keeping the batch boiling.
5. Boil down quickly to 69 deg. Brix (223 deg. F).
6. Take off steam, remove foam.
7. Add 300 cc citric acid solution 50%.
8. Fill hot (180 deg. F).
9. Invert receptacles for 3 minutes.

Check each batch for Brix 68-70 deg.; acidity/pH = 3.2 +/- 0.2.

Evaluate: absence of defects; colour; flavour; consistency.

Various fruit: sugar ratios can be manufactured; some basic recipes are as follows:

	Ratio 50:50	Ratio 45:55	Ratio 40:60
Fruit	55 lb	49.5 lb	44 lb
Water	-	11 lb	13.2 lb
Sugar	55 lb	60.5 lb	66 lb
Pectin (150 grade)	225 g	237.5 g	250 g
Citric acid (50% sol.)	300 cc	320 cc	335 cc

Fruit Juice Technologies

Fruit juices are products for direct consumption and are obtained by the extraction of cellular juice from fruit, this operation can be done by pressing or by diffusion.

For the purpose of this document, the technology of fruit juice processing will cover two finished product categories:

- juices without pulp ("clarified" or "not clarified");
- juices with pulp ("nectars").

We will also define as

a. "natural juices" products obtained from one fruit; and
b. "mixed juices" products obtained from the mix of two or three juices from different fruit species or by adding sugar.

Juices obtained by removal of a major part of their water content by vacuum evaporation or fractional freezing will be defined as "concentrated juices".

Technological Steps for Processing of Fruit Juices Without Pulp

Fruit juices must be prepared from sound, mature fruit only.

Soft fruit varieties such as grapes, tomatoes and peaches should only be transported in clean boxes which are free from mould and bits of rotten fruit.

Washing: fruit must be thoroughly washed. Generally, fruit will be submitted to a pre-washing before sorting and a washing step just after sorting.

Sorting: removal of partially or completely decayed fruit is the most important operation in the preparation of fruit for production of first quality fruit juices; sorting is carried out on moving inspection belts or sorting tables.

Crushing/Grinding/Disintegration Step is applied in different ways and depends on fruit types:

Crushing for grapes and berries;

Grinding for apples, pears;

Disintegration for tomatoes, peaches, mangoes, apricots.

This processing step will need specific equipment which differs from one type of operation to another.

Enzyme Treatment of crushed fruit mass is applied to some fruits by adding 2-8% pectolitic enzymes at about 50° C for 30 minutes.

This optional step has the following advantages: extraction yield will be improved, the juice colour is better fixed and finished product taste is improved.

However, for fruit which is naturally rich in pectic substances, this treatment makes the resulting "exhausted" material useless for industrial pectin production.

Heating of crushed fruit mass before juice extraction is an optional step used for some fruit in order to facilitate pressing and colour fixing; at same time, protein coagulation takes place.

Pressing to extract juice.

Diffusion is an alternative step for juice extraction and can be carried out discontinuously or in batteries at water temperature of about 80-85 ° C.

Juice Clarifying can be performed by centrifugation or by enzyme treatment. Centrifugation achieves a separation of particles in suspension in the juice and can be considered as a pre-clarifying step. This operation is carried out in centrifugal separators with a speed of 6000 to 6500 RPM.

Enzyme clarifying is based on pectic substance hydrolysis; this will decrease the juices' viscosity and facilitate their filtration. The treatment is the addition of pectolitic enzyme preparations in a quantity of 0.5 to 2 g/l and will last 2 to 6 hours at room temperature, or less than 2 hours at 50° C, a temperature that must not be exceeded.

The control of this operation is done by checking the decrease in juice viscosity. Sometimes, the enzyme clarifying is completed with the step called "sticking" by the addition of 5-8 g/hl of food grade gelatine which generates a flocculation of particles in suspension by the action of tannins.

Filtration of clarified juice can be carried out with kieselgur and bentonite as filtration additive in press-filters (equipment).

De-tartarisation is applied only to raisin juice and is aimed to eliminate potassium bitartrate from solution. This step can be performed by the addition of 1% calcium lactate or 0.4% calcium carbonate. Pasteurization of juice can be done for temporary preservation (pre-pasteurization) and in this case this operation is carried out with continuous equipment (heat exchangers, etc.); warm juice is stored in drums or large size receptacles (20-30 kg). Pasteurization conditions are at 75°C in continuous stream.

Pasteurization of bottled juice is then carried out just before delivery to the market; this is performed in water baths at 75° C until the point where the juice reaches 68° C. In cases when the final pasteurization is done without pre-pasteurization and temporary storage, modern methods use a rapid pasteurization followed by aseptic filling in receptacles. Rapid pasteurization conditions are as follows: temperature about 80° C, over 10-60 sec., followed by cooling; all operations are carried out in continuous stream. Preservation under CO2 pressure may be done at a concentration of 1.5% CO2 under a pressure of 7 kg/cm^2. At the distribution step, proceed at CO2 decompression and the juice is then submitted to a sterilising filtration and aseptic filling in receptacles.

Preservation by freezing is carried out at about -30° C, after a preliminary de-aeration; storage is at -15 to -20° C. Production of concentrated juices by evaporation is performed under vacuum (less than 100 mm Hg residual pressure) up to a concentration of 65-70% total sugar which assures preservation without further pasteurization. Modern evaporation installations recover flavours from juices which are then reincorporated in concentrated juices.

Additional operations for juice manufacturing are the vacuum de-aeration and mixing with other fruit juices or with sugar. For the production of non clarified juices the centrifugation is the only specific step, enzyme clarifying and subsequent filtration being eliminated. The optimum sugar/acid ratio for the majority of fruit, mainly for pomaces, is 10/1 to 15/1. Fruit which is rich in carotenoids (apricots, peaches, etc.) is only processed as juices with pulp ("nectars").

Technological Flow-sheet for Fruit Juices with Pulp ("Nectars")

This process is divided at industrial scale in two categories of operations:

a. processing for obtaining juices;

b. juice conditioning for preservation.

Operations in the first category are differ according to the type of fruit which to be processed. Pomaces (apples, pears) are washed and sorted and then crushed in a colloid mill; fruit purée is then passed through a screw type heating equipment where direct steam is used as a source of heat. Warm fruit mass is treated in a pulper with a 2 mm screen and then through an extractor similar with the equipment used for tomato juice.

Stone fruits (apricots, peaches, cherries, etc.) after washing and sorting are submitted to steam in a continuous heater, then the warm fruit mass is passed through a pulper and then an extractor (as mentioned above). Berries (strawberry, wild berries, etc.) are washed, sorted and then crushed, preheated and then introduced in extractor.

In order to avoid browning and undesirable taste modifications it is usual to add about 0.05% ascorbic acid. Second category type of operations are similar for all fruit species. Partial elimination of cellulose is achieved with a continuous centrifugal separator; the resulting juice is then processed in order to adjust sugar and acid content for viscosity. Sugar (about 8-10%) is added as a syrup (in water or in the juice of same fruit obtained by pressure). Acidity is adjusted with citric or tartaric acid. The adjusted juice is then deaerated under vacuum at about 40° C. This step aims at avoiding oxidative reactions and vitamin C loss reduction.

An important subsequent step is an intensive homogenisation (under pressure at 150-180 A) in order to obtain particles with dimensions below 100. The homogenised juice obtained is then continuously pasteurized in plate heat exchanger equipment at a temperature of about 130° C, cooled down to about 90° C and aseptically packed in receptacles. The principal characteristics of fruit "nectars" are uniformity and stability of the content provided by the advanced disintegration of fruits. Stability can be obtained by increasing product viscosity by adding pectin for fruit which is deficient in this component. In order to avoid "separation", intensive homogenisation is carried out as described above.

Fruit "nectars" contain all the important components of the original fruit and to a large extent maintain their taste and flavour. The sugar/ acidity (as citric acid) ratio is to a large extent determined by the type of fruit and the correction applied; for example, this ratio is 30 for apricots, 40 for peaches, 160 for pears, etc.

3

Enzymes in Milk and Cheese Industries

We have been using enzymes for thousands of years to make various drinks and things to eat, such as cheese, yoghurt, bread, beer, and wine. Of course the people who first discovered how to make these knew nothing about enzymes. In this unit we can see how these amazing biological catalysts are used in making these traditional foodstuffs, and also in making new products.

In the field of biotechnology there are many industrial applications that result in biotech products that we use everyday at home. Some of these are food science applications that utilize enzymes to produce or make improvements in the quality of different foods. In the dairy industry, some enzymes are required for the production of cheeses, yogurt and other dairy products, while others are used in a more specialized fashion to improve texture or flavour. Five of the more common types of enzymes and their role in the dairy industry are described below.

Rennet

Milk contains proteins, specifically caseins, that maintain its liquid form. Proteases are enzymes that are added to milk during cheese production, to hydrolyze caseins, specifically kappa casein, which stabilizes micelle formation preventing coagulation. Rennet and rennin are general terms for any enzyme used to coagulate milk. Technically rennet is also the term for the lining of a calf's fourth stomach. The most common enzyme isolated from rennet is chymosin. Chymosin can also be obtained from several other animal, microbial or vegetable sources, but indigenous microbial chymosin (from fungi or bacteria) is ineffective for making cheddar and other hard cheeses. Limited supplies of calf rennet have prompted genetic engineering of microbial chymosin by cloning calf prochymosin genes into bacteria. Bioengineered chymosin may be involved in production of up to 70% of cheese products. While use of bioengineered enzymes spares the lives of calves, it presents ethics issues for those opposed to eating foods prepared with GEMs.

Other Proteases

Milk contains a number of different types of proteins, in addition to the caseins. Cow milk also contains whey proteins such as lactalbumin and lactoglobulin. The denaturing of these whey proteins, using proteases, results in a creamier yogurt product. Destruction of whey proteins is also essential for cheese production.

During production of soft cheeses, whey is separated from the milk after curdling, and may be sold as a nutrient supplement for body building, weight loss, and lowing blood pressure, among other things. There have even been reports of dietary whey for cancer therapies, and having a role in the induction of insulin production for those with Type 2 diabetes. Proteases are used to produce hydrolyzed whey protein, which is whey protein broken down into shorter polypeptide sequences. Hydrolyzed whey is less likely to cause allergic reactions and is used to prepare supplements for infant formulas and medical uses.

Lactase

Lactase is a glycoside hydrolase enzyme that cuts lactose into it's constituent sugars, galactose and glucose. Without sufficient production of lactase enzyme in the small intestine, humans become lactose intolerant, resulting in discomfort (cramps, gas and diarrhea) in the digestive tract upon ingestion of milk products. Lactase is used commercially to prepare lactose-free products, particularly milk, for such individuals. It is also used in preparation of ice cream, to make a creamier and sweeter-tasting product. Lactase is usually prepared from *Kluyveromyces* sp. of yeast and *Aspergillus* sp. of fungi.

Catalase

The enzyme Catalase has found limited use in one particular area of cheese production. Hydrogen peroxide is a potent oxidizer and toxic to cells. It is used instead of pasteurization, when making certain cheeses such as Swiss, in order to preserve natural milk enzymes that are beneficial to the end product and flavour development of the cheese. These enzymes would be destroyed by the high heat of pasteurization.

However, residues of hydrogen peroxide in the milk will inhibit the bacterial cultures that are required for the actual cheese production, so all traces of it must be removed. Catalase enzymes are typically obtained from bovine livers or microbial sources, and are added to convert the hydrogen peroxide to water and molecular oxygen.

Lipases

Lipases are used to break down milk fats and give characteristic flavours to cheeses. Stronger flavoured cheeses, for example, the italian cheese, Romano, are prepared using lipases. The flavour comes from the free fatty acids produced when milk fats are hydrolyzed. Animal lipases are obtained from kid, calf and lamb, while microbial lipase is derived by fermentation with the fungal species *Mucor meihei.*

Although microbial lipases are available for cheese-making, they are less specific in what fats they hydrolyze, while the animal enzymes are more partial to short and medium-length fats. Hydrolysis of the shorter fats is preferred because it results in the desirable taste of many cheeses. Hydrolysis of the longer chain fatty acids can result in either soapiness, or no flavour at all.

Enzymes in Baking Industry

For decades, enzymes such as malt and fungal alpha-amylases have been used in bread-making. Rapid advances in biotechnology have made a number of exciting new enzymes available for the baking industry. The importance of enzymes is likely to increase as consumers demand more natural products free of chemical additives. For example, enzymes can be used to replace potassium bromate, a chemical additive that has been banned in a number of countries.

The dough for bread, rolls, buns and similar products consists of flour, water, yeast, salt and possibly other ingredients such as sugar and fat. Flour consists of gluten, starch, non-starch polysaccharides, lipids and trace amounts of minerals. As soon as the dough is made, the yeast starts to work on the fermentable sugars, transforming them into alcohol and carbon dioxide, which makes the dough rise.

The main component of wheat flour is starch. Amylases can degrade starch and produce small dextrins for the yeast to act upon. There is also a special type of amylase that modifies starch during baking to give a significant anti-staling effect

Table: *Lists some of the bread properties that can be improved using industrial enzymes.*

Enzyme	*Effect*
Amylase	Maximises the fermentation process to obtain an even crumb structure and a high loaf volume
Maltogenic alpha-amylases	Improves shelf-life of bread and cakes
Glucose oxidase	Oxidative reaction with gluten to make weak doughs stronger, drier and more elastic
Lipase	Modifies the natural lipids in flour to strengthen the dough
Lipoxygenase	Bleaching and strengthening dough
Xylanase	Dough conditioning. Easier dough handling and improved crumb structure
Protease	Weakens the gluten to give plastic properties required in doughs for biscuits
Asparginase	Reduces the amount of acrylamid formed during baking

Gluten is a combination of proteins that forms a large network during dough formation. This network holds the gas in during dough proofing and baking. The strength of this gluten network is therefore extremely important for the quality of all bread raised using yeast. Enzymes such as hemicellulases, xylanases, lipases and oxidases can directly or indirectly improve the strength of the gluten network and so improve the quality of the finished bread.

Flour Supplementation

Malt flour and malt extract can be used as enzyme supplements because malt is rich in alpha-amylases. Commercial malt preparations can differ widely in their enzyme activity, whereas an industrial enzyme is supplied with a standardised activity.

The alpha-amylases degrade the damaged starch in wheat flour into small dextrins, which allows yeast to work continuously during dough fermentation, proofing and the early stage of baking. The result is improved bread volume and crumb texture. In addition, the small oligosaccharides and sugars such as glucose and maltose produced by these enzymes enhance the Maillard reactions responsible for the browning of the crust and the development of an attractive baked flavour.

Bread and cake staling is responsible for significant financial loss for both consumers and producers. For instance, every year in the

USA bread worth more than USD 1 billion is discarded. However the main saving on prolonging the shelf life is actually savings in transportation and fuel costs due to a more efficient distribution. Staling is associated with a loss of freshness in terms of increased crumb firmness, decreased crumb elasticity and loss of moistness.

Staling is believed to be due to changes in starch structure during storage. When the starch granules revert from a soluble to an insoluble form, they lose their flexibility; the crumb becomes hard and brittle. For decades, emulsifiers have been used as anti-staling agents. However, they actually have a limited anti-staling effect and are subject to special labelling rules.

By contrast, Novozymes' bacterial maltogenic alpha-amylases has been found to have a significant anti-staling effect. It modifies the starch during baking at the temperature when most of the starch starts to gelatinise. The resulting modified starch granules remain more flexible during storage. Bread produced with maltogenic alpha-amylase has a far softer and more elastic crumb than bread produced with distilled monoglycerides as emulsifiers.

Dough Conditioning

Flour contains 2.5-3.5% non-starch polysaccharides, which are large polymers (mainly pentosans) that play an important role in bread quality due to their water absorption capability and interactions with gluten. Although the true mechanism of hemicellulase, pentosanase or xylanase in bread-making has not been clearly demonstrated, it is well known that the addition of certain types of pentosanases or xylanases at the correct dosage can improve dough machinability yielding a more flexible, easier-to-handle dough. Consequently, the dough is more stable and gives better ovenspring during baking, resulting in a larger volume and improved crumb texture.

Normal wheat flour contains 1-1.5% lipids, both polar and non-polar. Some of these lipids, especially the polar lipids such as phosphorlipids and galactolipids are able to stabilise the air bubbles in the gluten matrix. The addition of functional lipases modifies the natural flour lipids so they become better at stabilizing the dough. This ensures a more stable dough in case of over-fermentation, a larger loaf volume, and significantly improved crumb structure. Because of the more uniform and smaller crumb cells, the crumb texture is silkier and the crumb colour appears to be whiter. It also reduces the need for addition of emulsifiers like DATEM and SSL that otherwise are commonly added to dough in order to stabilise it. This in turn means that emulsifiers can be removed from the label.

Chemical oxidants such as bromates, azodicarbonamide and ascorbic acid have been widely used to strengthen the gluten when making bread. As an alternative, oxidases such as glucose oxidase can partially replace the use of these chemical oxidants and achieve better bread quality.

The Synergistic Effects of Enzymes

Each of the enzymes mentioned above has its own specific substrate in wheat flour dough. For example, lipases work on the lipids, xylanase works on the pentosans, and amylases work on the starch. Because the interaction of these substrates in dough and bread is rather complex, the use of enzyme combinations can have synergistic effects that are not seen if only one enzyme is used, not even at high dosages. Quite often an overdosage of enzymes will have a detrimental effect on either the dough or the bread.

For instance, an overdose of fungal alpha-amylase or hemicellulase/xylanase may result in a dough that is too sticky to be handled by the baker or baking equipment. It is therefore beneficial for some types of bread formulation to use a combination of lower dosages of alpha-amylase and xylanase with low dosages of lipase or glucose oxidase to achieve optimum dough consistency stability and bread quality. Another example is to use maltogenic alpha-amylase in combination with fungal alpha-amylases and xylanase or lipase to secure optimum crumb softness as well as optimum bread quality in terms of crumb structure, bread volume, etc.

Reduction of Acrylamide Content in Food Products

During recent years it has been shown that the amount of the potentially carcinogenic substance acrylamide is relatively high in a number of cereal and potato based products like biscuits, crackers, crisp bread, French fries and potato crisps. This is a substance that is formed at high temperatures when the amino acid asparagine reacts with a reducing sugar like glucose. To meet this issue the enzyme asparaginase have been developed in order to reduce the formation of acrylamide. Asparaginase converts asparagine to aspartic acid which does not take part in the formation of acrylamide. The use of an asparginase is able to reduce the formation of acrylamide with up to 90%.

Enzymes in Fruit Juices and Beverages Industry

Enzyme Used in Fruit Juice Manufacturing

Enzymes are, widely used in fruit Juice and brewing industries to achieve specific objectives, which can not be otherwise, achieved. In addition, brewing industry employs varied but specific strains of yeast, e.g., S. cerevisiae, S. carisbergensis, etc., for, fermentation of the brew.

The cloudiness of fruit juices and wines is mainly due to pectins, which exhibit are usually associated with other plant polymers and even cell debris. The pectins are digested by pecteolytic enzymes prepared from A. niger. These enzyme preparations are mixtures of the following:

(i) polygalacturonase (random digestion of pectin),

(ii) pectin esterase (removal of methyl esters and release of methanol; pectin become degradable by polygalacturonase),

(iii) pectin lyase (cleaves pectins into oligosaccharides even without pectin esterase action), and

(iv) hemicellulase (a mixture of enzymes degrading hemicellulose). They are generally added directly into the fruit pulp at about 20 U/l.

The four enzymes present in the mixture act synergistically to accomplish, a task, which can not be achieved by mechanical means. Treatment of fruit pulp with pecteolytic enzyme mixture gives the following benefits:

(i) Elimination of juice/wine cloudiness,

(ii) reduced solution viscosity,

(iii) Increased juice yields, e.g., a 15% increase in case of white grapes, and

(iv) Shorter fermentation period in case of wine making. In addition,

(v) pectins stabilize the cell debris in a colloidal state; but once pectins are digested, the debris precipitate and are removed by filtration.

Uses of Enzymes in Fruit Juice Wine Brewing Distilling Industries

One of the major problems in the preparation of fruit juices and wine is cloudiness due primarily to the presence of pectins. These consist primarily of a-1, 4-anhydrogalacturonic acid polymers, with varying degrees of methyl esterification. They are associated with other plant polymers and, after homogenization, with the cell debris. The cloudiness that they cause is difficult to remove except by enzymic hydrolysis. Such treatment also has the additional benefits of reducing the solution viscosity, increasing the volume of juice produce (e.g. the yield of juice from white grapes can be raised by 15%), subtle but generally beneficial changes in the flavour and, in the case of wine-making, shorter fermentation times.

The enzymes used in brewing are needed for saccharification of starch (bacterial and fungal a-amylases), breakdown of barley a-1.4- and a-1,3-linked glucan (b-glucanase) and hydrolysis of protein (neutral protease) t increase the (later) fermentation rate, particularly in the production of high-gravity beer, where extra protein is added. Cellulases are also occasionally used, particularly where wheat is used as adjunct to help break down the barley is b-glucans.

Due to the extreme heat stability of the B. amyloliquefaciens a-amylase where this is used, the wort must be boiled for a much longer period (e.g. 30 minutes) to inactivate it prior to fermentation. Papain is used in the later post-fermentation stages of beer-making to prevent the occurrence of protein-and tannin- containing 'chill-haze' otherwise formed on cooling the beer.

Recently, 'light' beers, of lower calorific content, have become more popular. These require a higher degree of saccharification at lower starch concentrations to reduce the alcohol and total solid content of the beer. This may be achieved by the use of glucoamylase and/or fungal a-amylase during the fermentation.

A new enzyme preparation of fungal pectin lyase (EC 4.2.2.10) was shown to be useful for the production of cranberry juice and clarification of apple juice in the food industry (Semenova et al., 2006). A comparative study showed that the preparation of pectin lyase is competitive with commercial pectinase products. The molecular weight of homogeneous pectin lyase was 38 kDa. Properties of the homogenous enzyme were studies. This enzyme was most efficient in removing highly esterified pectin.

The zygomycete microfungus R. microsporus var. microsporus produced a 1,3-1,4-beta-D-glucan 4-glucanhydrolase (EC 3.21.73) which was able to hydrolyse beta-D-glucan that contains both the 1,3-and 1,4-bonds (barley beta-glucans) (Celestino et al., 2006). Its molecular mass was 33.7 kDa.

Maximum activity was detected at pH values in the range of 4-5, and temperatures in the range of 50-60°C. The enzyme was able to reduce both the viscosity of the brewermash and the filtration time, indicating its potential value for the brewing industry Landbo et al, (2006) examined the clarification and haze-diminishing effects of alternative clarification strategies on black current juice including centrifugation and addition of acidic protease and pectinolytic enzyme preparation and gallic acid.

The extent of clarification and haze diminishment varied after individual treatment with five different acidic proteases, but one of the protease preparations, Enzeco, derived from Aspergillus niger, consistently tended to perform best.

The individual and interactive effects on juice turbidity, total phenols, and total anthocyanin content of clarification treatment involving the use of two selected acid proteases (Enzeco and Novozyme 89L), a pectinase (Pectinex BE 3-L), and gallic acid were evaluated in a full factorial 2(4) experimental design. Haze development during cold storage decreases when gallic acid or any of the enzyme preparation was employed individually, but negative interaction effects resulted when the pectinase was employed in combination with any of the proteases. After 28 storage days at 2°C, the lowest levels of haze formation were achieved when the Enzeco protease preparation, added at 0.025 g/L, was added with 0.050 g/L of gallic acid and allowed to react in the juice for 90 min. at 50°C. The corresponding anthocyanin reduction was approximately 12% (compared to approximately 30% with gelatin silica soltreatment).

The data support the hypothesis that phenol-protein interactions are involved in juice turbidity development during cold storage of berry juices and demonstrate that pre-centrifugation and protease-assisted clarification show promise as an alternative, phenolic-retaining clarification strategy in black currant juice processing.

At monosodium glutamate (MSG) biosensor with immobilized L-glutamate oxidase (L-GLOD) has been developed and studied for analysis of MSG in sauces, soups, etc. (Basu et al., 2006). The immobilized enzymatic membrane was attached with oxygen electrode with a push-cap system. The detection limit of the sensor was 1 mg/dl and the standard curve was found to be linear up to 20 mg/dl. Response time of the sensor was 2 minutes. Cross-linking with glutaraldehyde in the presence of bovine serum albumin (BSA) as a spacer molecule has been used for immobilization.

Optimization of the sensor was done with an increase in L-GOLD concentration (6.3-31.5 IU)(and also with increase in loading volume of enzyme solution (5-20 microL). Optimization of pH and temperature was also studied. The permeability of O2 through different membrane was studied with and without immobolized L-GLOD. The enzymatic membrane was used for 20 measurements and stability of membrane was observed.

Enzymes in Meat Industries

Papain

Papain, also known as papaya proteinase I, is a cysteine protease (EC 3.4.22.2) enzyme present in papaya (*Carica papaya*) and mountain papaya (*Vasconcellea cundinamarcensis*).

Papain Family

Papain belongs to a family of related proteins with a wide variety of activities, including endopeptidases, aminopeptidases, dipeptidyl peptidases and enzymes with both exo- and endo-peptidase activity. Members of the papain family are widespread, found in baculovirus, eubacteria, yeast, and practically all protozoa, plants and mammals. The proteins are typically lysosomal or secreted, and proteolytic cleavage of the propeptide is required for enzyme activation, although bleomycin hydrolase is cytosolic in fungi and mammals.

Papain-like cysteine proteinases are essentially synthesised as inactive proenzymes (zymogens) with N-terminal propeptide regions. The activation process of these enzymes includes the removal of propeptide regions. The propeptide regions serve a variety of functions in vivo and in vitro.

The pro-region is required for the proper folding of the newly synthesised enzyme, the inactivation of the peptidase domain and stabilisation of the enzyme against denaturing at neutral to alkaline pH conditions. Amino acid residues within the pro-region mediate their membrane association, and play a role in the transport of the proenzyme to lysosomes. Among the most notable features of propeptides is their ability to inhibit the activity of their cognate enzymes and that certain propeptides exhibit high selectivity for inhibition of the peptidases from which they originate.

Structure

The papain precursor protein contains 345aa, and consists of a signal sequence (1-18), a propeptide (19-133) and the mature peptide (134-345). The amino acid numbers are based on the mature peptide. The protein is stabilised by 3 disulfide bridges. Its 3D structure consists of 2 distinct structural domains with a cleft between them. This cleft contains the active site, which contains a catalytic diad that has been likened to the catalytic triad of chymotrypsin. The catalytic diad is made up of the amino acids - cysteine-25 (from which it gets its classification) and histidine-159. Aspartate-158 was thought to play a role analogous to the role of aspartate in the serine protease catalytic triad, but that has since then been disproved. None-the-less, it does appear to play a role.

Function

The mechanism by which it breaks peptide bonds involves deprotonation of Cys-25 by His-159. Asparagine-175 helps to orient the imidazole ring of His-159 to allow this deprotonation to take place. Cys-25 then performs a nucleophilic attack on the carbonyl carbon of a peptide backbone. This frees the amino terminal of the peptide, and forms a covalent acyl-enzyme intermediate. The enzyme is then deacylated by a water molecule, and releases the carboxy terminal portion of the peptide. In immunology, papain is known to cleave the Fc (crystallisable) portion of immunoglobulins (antibodies) from the Fab (antigen-binding) portion.

Uses

Its utility is in breaking down tough meat fibers and has been utilized for thousands of years in its native South America. It is sold as a component in powdered meat tenderizer available in most supermarkets. Papain, in the form of a meat tenderizer such as Adolph's, made into a paste with water, is also a home remedy treatment for jellyfish, bee, and yellow jacket (wasps) stings; mosquito bites; and possibly stingray wounds, breaking down the protein toxins in the venom. Papain can be used to dissociate cells in the first step of cell culture preparations. A ten-minute treatment of small tissue pieces (less than 1 mm cubed) will allow papain to begin breaking down the extracellular matrix molecules holding the cells together. After ten minutes, the tissue should be treated with a protease inhibitor solution to stop the protease action (if left untreated, papain's activity will lead to complete lysis of the cells). The tissue must then be triturated (passed quickly up and down through a Pasteur pipette) in order to break up the pieces of tissue into a single cell suspension.

It is also used as an ingredient in various enzymatic debriding preparations, notably Accuzyme. These are used in the care of some chronic wounds to clean up dead tissue.

Papain can also be found as an ingredient in some toothpastes or mints as teeth-whitener. Its whitening effect in toothpastes and mints is minimal, however, because the papain is present in low concentrations, and is quickly diluted by saliva. It would take several months of using the whitening product to have noticeably whiter teeth. It is the main ingredient of Papacarie, a gel used for chemomechanical dental caries removal. Besides the advantage of avoiding the use of rotary cutting tools, it does not interfere in the bond strength of restorative materials to dentin.

Immunoglobulins

An antibody digested by papain yields three fragments: two 50 kDa Fab fragments and one 50kDa Fc fragment. The papain-digested antibody is unable to promote agglutination, precipitation, opsonization, and lysis.

Bromelain

Bromelain can refer to one of two protease enzymes extracted from the plant family Bromeliaceae, or it can refer to a combination of those enzymes along with other compounds produced in an extract.

Extract Components

Bromelain extract is a mixture of protein-digesting enzymes— called proteolytic enzymes or proteases and several other substances in smaller quantities. The proteolytic enzymes are referred to as sulfhydryl proteases since a free sulfhydryl group of a cysteine side-chain is required for function. The two main enzymes are:

- Stem bromelain - EC 3.4.22.32
- Fruit bromelain - EC 3.4.22.33

The other substances typically include peroxidase, acid phosphatase, protease inhibitors, and calcium.

History

The first isolation of bromelain was recorded by the Venezuelan chemist Vicente Marcano (BU1 1.Phar. 5,77) in 1891 from the fruit of pineapple. In 1892, Chittenden, assisted by Joslin and Meara, investigated the matter fully (Trans. Conn. Acad. Arts Sci. 8, 281-308), and called it 'bromelin'. Later the term 'bromelain' was introduced and originally applied to any protease from any plant member of the plant family Bromeliaceae. Bromelain was first introduced as a therapeutic supplement in 1957. Research on bromelain apparently was first conducted in Hawaii but more recently has been conducted in countries in Asia, Europe and Latin America. Germany has recently taken a great interest in bromelain research; bromelain is currently the 13th most widely used herbal medicine in Germany.

Source

Bromelain is present in all parts of the pineapple plant (*Ananas sp.*), but the stem is the most common commercial source, presumably because it is readily available after the fruit has been harvested. Pineapples have had a long tradition as a medicinal plant among the natives of South and Central America.

Uses

Meat Tenderizing

Along with papain, bromelain is one of the most popular substances to use for meat tenderizing. Today, about 90% of meat tenderizer is used in consumer households. Bromelain is sold in a powdered form, which is combined with a marinade, or directly sprinkled on the uncooked meat. The enzyme will penetrate the meat, and by a process called forking, cause the meat to be tender and palatable when cooked. If the enzyme is allowed to work for too long, the meat may become too "mushy" for many consumers' preferences. Cooked or canned pineapple does not have a tenderizing effect, as the enzymes are heat labile.

Some prepared meat products, such as meatballs, and commercially available marinades include pineapple and/or pineapple-derived ingredients. Apart from the mushiness of meat that has been over-tenderized, the activity of bromelain and similarly proteolytic plant enzymes can be undesirable where it is inappropriate. In dishes that depend on their protein content for important attributes, uncooked pineapple or its juice can be a nuisance. For example, some dishes such as brawn and jelly [called gelatin dessert (Jell-o) in the US] rely on the setting of gelatin.

They will not set if they contain raw pineapple or pineapple juice. Raw figs, papaya, and similarly proteolytic vegetable matter causes similar problems. To prevent the problem, the fruit should first be cooked, or at least parboiled sufficiently to inactivate the enzymes. Thorough heating to above about 65 degrees C will generally suffice. This problem does not affect dishes based on nonproteinaceaous gelling agents, such as agar, though such agents may be prevented from gelling by too much heating in acid recipes.

4

Food Preservation Technology

Food preservation is the process of treating and handling food to stop or slow down spoilage (loss of quality, edibility or nutritional value). Preservation usually involves preventing the growth of bacteria, yeasts, fungi, and other micro-organisms (although some methods work by introducing benign bacteria, or fungi to the food), as well as retarding the oxidation of fats which cause rancidity. Food preservation also includes processes which inhibit natural discoloration that can occur during food preparation, such as the enzymatic browning reaction in apples after they are cut.

Many processes designed to preserve food will involve a number of food preservation methods. Preserving fruit, by turning it into jam, for example, involves boiling (to reduce the fruit's moisture content and to kill bacteria, yeasts, etc.), sugaring (to prevent their re-growth) and sealing within an airtight jar (to prevent recontamination). There are many traditional methods of preserving food that limit the energy inputs and reduce carbon footprint.

Maintaining or creating nutritional value, texture and flavour is an important aspect of food preservation, although, historically, some methods drastically altered the character of the food being preserved. In many cases these changes have now come to be seen as desirable qualities – cheese, yoghurt and pickled onions being common examples.

Preservation Processes

Preservation processes include:

- Heating to kill or denature micro-organisms (e.g. boiling)
- Oxidation (e.g. use of sulphur dioxide)

- Toxic inhibition (e.g. smoking, use of carbon dioxide, vinegar, alcohol etc.)
- Dehydration (drying)
- Osmotic inhibition (e.g. use of syrups)
- Low temperature inactivation (e.g. freezing)
- Ultra high water pressure (e.g. fresherized, a kind of "cold" pasteurization, the pressure kills naturally occurring pathogens, which cause food deterioration and affect food safety.)
- Combinations of these methods.

Drying

Drying is a method of food preservation that works by removing water from the food, which inhibits the growth of microorganisms and hinders quality decay. Drying food using sun and wind to prevent spoilage has been practised since ancient times. Water is usually removed by evaporation (air drying, sun drying, smoking or wind drying) but, in the case of freeze-drying, food is first frozen and then the water is removed by sublimation.

Bacteria yeasts and moulds need the water in the food to grow. Drying effectively prevents them from surviving in the food.

Food Types

Many different foods are prepared by dehydration. Good examples are meat such as prosciutto (a.k.a. Parma ham), bresaola, and beef jerky. Dried and salted reindeer meat is a traditional Sami food. First the meat is soaked / pickled in saltwater for a couple of days to guarantee the conservation of the meat. Then the meat is dried in the sun in spring when the air temperature is below zero. The dried meat can be further processed to make soup. Fruits change character completely when dried: the plum becomes a prune, the grape a raisin; figs and dates are also transformed in new, different products, that can be eaten as they are or else after rehydration.

Home drying of vegetables, fruit and even meat (to produce jerky) may be carried out by a do-it-yourself practice, employing electrical dehydrators (household appliance). If the user does not like to use additives as potassium metabisulphite, or BHA, BHT for meats, dried products may be hermetically shelf stored if it is to be consumed soon, or else in the refrigerator or even freezer if a long storage is to be expected. Freeze dried vegetables are often found in backpackers food, hunters, military, etc. The exception to this rule are bulbs, such as garlic and onion, which are often dried. Also chilis are frequently dried. Edible and psilocybin mushrooms, as well as other fungi, are also sometimes dried for preservation purposes, to affect the potency of chemical components, or so they can be used as seasonings.

For centuries, much of the European diet depended on dried cod, known as salt cod or bacalhau (with salt) or stockfish (without). It formed the main protein source for the slaves on the West Indian plantations, and was a major economic force within the triangular trade. Dried shark meat, known as Hαkarl, is a delicacy in Iceland.

Grain Drying

Hundreds of millions of tonnes of wheat, corn, soybean, rice and other grains as sorghum, sunflower seeds, rapeseed/canola, barley, oats, etc., are dried in grain dryers. In the main agricultural countries, drying comprises the reduction of moisture from about 17-30%w/w to values between 8 and 15%w/w, depending on the grain. The final moisture content for drying must be adequate for storage. The more oil the grain has, the lower its storage moisture content will be (though its initial moisture for drying will also be lower). Cereals are often dried to 14% w/w, while oilseeds, to 12.5% (soybeans), 8% (sunflower) and 9% (peanuts). Drying is carried out as a requisite for safe storage, in order to inhibit microbial growth. However, low temperatures in storage are also highly recommended to avoid degradative reactions and, especially, the growth of insects and mites. A good maximum storage temperature is about 18°C.

The largest dryers are normally used "Off-farm", in elevators, and are of the continuous type: Mixed-flow dryers are preferred in Europe, while Cross-flow dryers in the USA. In Argentina, both types are usually found. Continuous flow dryers may produce up to 100 metric tonnes of dried grain per hour. The depth of grain the air must traverse in continuous dryers range from some 0.15 m in Mixed flow dryers to some 0.30 m in Cross-Flow. Batch dryers are mainly used "On-Farm", particularly in the USA and Europe. They normally consist of a bin, with heated air flowing horizontally from an internal cylinder through an inner perforated metal sheet, then through a annular grain bed, some 0.50 m thick (coaxial with the internal cylinder) in radial direction, and finally across the outer perforated metal sheet, before being discharged to the atmosphere. The usual drying times range from 1 h to 4 h depending on how much water must be removed, type of grain, air temperature and the grain depth. In the USA, continuous counterflow dryers may be found on-farm, adapting a bin to slowly drying grain fed at the top and removed at the bottom of the bin by a sweeping auger. Grain drying is an active area of manufacturing and research. Now it is possible to simulate the performance of a dryer with computer programs based on equations (mathematical models) that represent the phenomena involved in drying: physics, physical chemistry, thermodynamics and heat and mass transfer. Most recently the evolution of quality indices is beginning to be predicted with some confidence, in order to add an essential performance parameter with which to establish a compromise of reasonably fast drying rate, limited energy consumption, and satisfactory grain quality. A typical quality parameter in wheat drying is the breadmaking quality and germination percentage whose reductions in drying are somewhat related.

Methods

There are many different methods for drying, each with their own advantages for particular applications; these include:

- Bed dryers
- Drum drying
- Freeze Drying
- Shelf dryers
- Spray drying
- Sunlight
- Commercial food dehydrators
- Household oven.

One of the oldest methods of food preservation is by drying, which reduces water activity sufficiently to prevent or delay bacterial growth.

Refrigeration

Refrigeration preserves food by slowing down the growth and reproduction of micro-organisms and the action of enzymes which cause food to rot. The introduction of commercial and domestic refrigerators drastically improved the diets of many in the Western world by allowing foods such as fresh fruit, salads and dairy products to be stored safely for longer periods, particularly during warm weather.

Freezing

Freezing food preserves food from the time it is prepared to the time it is eaten. Since early times, farmers, fishermen, and trappers have preserved their game in unheated buildings during the winter season.

Freezing food slows down decomposition by turning water to ice, making it unavailable for most bacterial growth. In the food commodity industry, the process is called IQF or Individually Quick Frozen.

Process

History

Clarence Birdseye, an American inventor, developed the quick-freezing system. He discovered that the combination of ice, wind, and low temperatures in the Arctic froze anything that was exposed to it almost instantly. Birdseye's soon realized that the quick freezing effectively prevented large ice crystals from forming. Other attempts had resulted in the formation of large crystals, which destroyed the delicate cellular structure of the food. With only an electric fan, a few buckets of brine, and cakes of ice, Clarence Birdseye perfected his system of packing fresh food into waxed cardboard boxes and flash-freezing it under high pressure. He sold the patent to the Goldman-Sachs Trading Corporation (a subsidiary of Goldman Sachs & Company) and the Postum Company. In 1929 the first quick-frozen vegetables were sold to the public.

Modern Techniques

Manufacturers freeze foods by immersing them in very cold liquids, liquid nitrogen being the preferred medium. Nitrogen liquefies at about -320 °F (-195.5 °C), making it useful for quickly freezing foods. When food is submerged in liquid nitrogen, it rapidly freezes. The faster food freezes, the smaller the crystals that form within it. High-Pressure Shift Freezing is another method used to manufacture frozen food. It uses the principles of water's phase diagram. At a very high pressure, 900MPa, ice may be formed at room temperature. This is not an efficient way to create frozen foods, but it is being researched for future use. Dehydrofreezing is a commercial method used to reduce the cost of shipping, handling, and storage of fruits and vegetables. During dehydrofreezing, food is first dehydrated to the desired moisture level and then frozen. Fruits and vegetables have a higher water content than meats, which makes them more susceptible to the formation of large ice crystals. Dehydrofreezing gives the manufacturer peace of mind and keeps produce fresher.

Preservatives

Frozen foods don't require many preservatives because the process of preparing the food for freezing kills much of the bacteria living on the food. Carboxymethylcellulose (CMC) is used as a stabilizer in frozen foods because of its tasteless and odorless properties.

Packaging

Frozen food packaging must maintain its integrity throughout machine filling, sealing, freezing, storage, transportation, thawing, and often cooking. Most frozen foods are cooked in a microwave oven. To make it easier for the consumer, manufacturers have developed packaging that can go straight from freezer to microwave.

In 1974, the first differential heating container (DHC) was sold to the public. A DHC is a sleeve of metal designed to allow frozen foods to receive the correct amount of heat. Various sized apertures were positioned around the sleeve. The consumer would put the frozen dinner into the sleeve according to what needed the most heat. This ensured proper cooking. Today there are multiple options for packaging frozen foods. Boxes, cartons, bags, pouches, heat-in-bag pouches, lidded trays and pans, crystallized PET trays, and composite and plastic cans.

Scientists are continually researching new aspects of frozen food packaging. Active packaging offers a host of new technologies that can actively sense and then neutralize the presence of bacteria or other harmful species. Active packaging can extend shelf-life, maintain product safety, and help preserve the food over a longer period of time. Several functions of active packaging are being researched:

- Oxygen scavengers
- Time Temperature Indicators and digital temperature dataloggers
- Antimicrobials
- Carbon Dioxide controllers
- Microwave susceptors
- Moisture control: Water activity, Moisture vapour transmission rate, etc.
- Flavour enhancers
- Oder generators
- Oxygen-permeable films
- Oxygen generators
- Validation of cold chain.

With these new technologies, food may last longer and our knowledge about its safety will increase.

Effects on Nutrients

Vitamin Content of Frozen Foods

Vitamin C: Usually lost in a higher concentration than any other vitamin. A study was performed on peas to determine the cause of Vitamin C loss. A vitamin loss of ten percent occurred during the blanching phase with the rest of the loss occurring during the cooling and washing stages. The vitamin loss was not actually accredited to the freezing process. Another experiment was performed involving peas and lima beans. Frozen and canned vegetables were both used in the experiment. The frozen vegetable were stored at -10 °F and the canned vegetables were stored at room temperature (75 °F). After 0, 3, 6, and 12 months of storage, the vegetables were analyzed with and without cooking. O'Hara, the scientist performing the experiment said, "From the view point of the vitamin content of the two vegetables when they were ready for the plate of the consumer, there did not appear to be any marked advantages attributable to method of preservation, frozen storage, processed in a tin, or processed in glass."

Vitamin B_1 (Thiamin): A vitamin loss of 25 percent is normal. Thiamin is easily soluble in water and is destroyed by heat.

Vitamin B_2 (Riboflavin): Not much research has been done to see how much freezing affects Riboflavin levels. One study found an 18 percent vitamin loss in green vegetables while another found a 4 percent loss. It is commonly accepted that the loss of Riboflavin has to do with the preparation for freezing rather than the actual freezing process itself.

Vitamin A (Carotene): There is little loss of Carotene during preparation for freezing and freezing of most vegetables. However, there is a danger of losing the vitamin during a long-continued storage period.

Efficiency

Freezing is an effective form of food preservation because the pathogens that cause food spoilage are killed or do not grow very rapidly at reduced temperatures. The process is less effective in food preservation than are thermal techniques, such as boiling, because pathogens are more likely to be able to survive cold temperatures rather than hot temperatures. One of the problems surrounding the use of freezing as a method of food preservation is the danger that pathogens deactivated (but not killed) by the process will once again become active when the frozen food thaws.

Foods may be preserved for several months by freezing. Long-term frozen storage requires a constant temperature of -18 °C (0 °F) or less. Some freezers cannot achieve such a low temperature. The time food can be kept in the freezer is reduced considerably if the temperature in a freezer fluctuates; small ice crystals thaw as the temperature moves up, and refreeze onto larger crystals as the temperature declines. Fluctuations can occur by a small gap in the freezer door or adding a large amount of unfrozen food.

Freezing is also one of the most commonly used processes commercially and domestically for preserving a very wide range of food including prepared food stuffs which would not have required freezing in their unprepared state. For example, potato waffles are stored in the freezer, but potatoes themselves require only a cool dark place to ensure many months' storage. Cold stores provide large volume, long-term storage for strategic food stocks held in case of national emergency in many countries.

Thermization

Thermization, also spelled thermisation, is a method of sterilizing raw milk with heat. The process is not used on other food products, and is similar to pasteurization but uses lower temperatures, allowing the milk product to retain more of its original taste. In Europe, there is a distinction between cheeses made of thermized milk and raw-milk cheeses. However, the United States' Food and Drug Administration (FDA) places the same regulations on all unpasteurized cheeses. As a result, cheeses from thermized milk must be aged for 60 days or more before being sold in the United States, the same restriction placed on raw-milk cheeses by the FDA.

Thermization involves heating milk at temperatures of around 145–149 °F (63–65 °C) for 15 seconds, while pasteurization involves heating milk at 160 °F (71 °C) for 15 seconds or at 145 °F (63 °C) for 30 minutes. Thermization is used to extend the keeping quality of raw milk (the length of time that milk is suitable for consumption) when it cannot be immediately used in other products, such as cheese. Thermization can also be used to extend the storage life of fermented milk products by inactivating microorganisms in the product.

Thermization inactivates psychrotrophic bacteria in milk, preventing the growth of heat-resistant enzymes and allowing the milk to be stored below 8 °C (46 °F) for three days, or stored at "-1 °C (30.2 °F) for seven days. Later, the milk may be given stronger heat treatment to be preserved longer. Cooling thermized milk before reheating it is necessary for the destruction of bacterial spores. When the milk is first heated, spores begin to germinate, but their growth is halted when the milk is refrigerated. The resulting vegetative bacteria are easier to kill with another heat treatment than the original spores.

Vacuum Packing

Vacuum packing is a method of storing food and presenting it for sale. Appropriate types of food are stored in an airless environment, usually in an air-tight pack or bottle to prevent the growth of microorganisms.

The vacuum environment removes atmospheric oxygen, protecting the food from spoiling by limiting the growth of aerobic bacteria or fungi, and preventing the evaporation of volatile components. Vacuum packing is commonly used for long-term storage of dry foods such as cereals, nuts, cured meats, cheese, smoked fish, coffee, and potato chips (crisps). It is also for storage of fresh foods such as vegetables, meats, and liquids such as soups in a shorter term because vacuum condition cannot stop bacteria from getting water which can promote their growth. Vacuum packaging food can extend its life by up to 3-5 times.

Vacuum packing is also used to reduce greatly the bulk of non-food items. For example, clothing and bedding can be stored in bags evacuated with a domestic vacuum cleaner or a dedicated vacuum sealer. This technique is sometimes used to compact household waste, for example where a charge is made for each full bag collected. Vacuum packing can be used to reduce bulk of inflatable items as well.

Vacuum packaging products using plastic bags, canisters, bottles, or mason jars are available for home use. Vacuum packaging delicate food items can be done by using an inert gas kit, typically available on chamber vacuum sealers. After air has been removed, an inert gas (such as nitrogen) is added to maintain the preservation of packaged food while preventing damage. An example of inert gas for packaging delicate foods is potato chips.

External Sealers

External vacuum sealers involve a bag being attached to the vacuum-sealing machine externally. The machine will remove the air and seal the bag, which is all done outside the machine.

Chamber Sealers

Chamber sealers require the entire product to be placed within the machine. Like external sealers, a plastic bag is typically used for packaging. Once the product is placed in the machine, the lid is closed and air is removed. Once the air is removed, the bag is sealed and the atmosphere within the chamber is returned back to normal. The lid is then opened and the product removed. Chamber sealers are typically used for higher-volume packaging. Manufacturers of chamber type vacuum packing machines include: Cryovac, Multivac, Sammic, VC999, New Diamond Vac , Jaw Feng Machinery ,Sevana and several others.

Preventing Freezer Burn

When foods are frozen without preparation, freezer burn can occur. It happens when the surface of the food is dehydrated, and this leads to a dried and leathery appearance. Freezer burn also ruins the flavour and texture of foods. Vacuum packing prevents freezer burn by preventing the food from exposure to the cold, dry air.

Sous-vide Cooking

Vacuum packaging also allows for a special cooking method, Sous-vide. Sous-vide, meaning "under vacuum" in French, involves poaching food that is vacuum sealed in a plastic bag.

Security

Due to an oxygen-poor environment, anaerobic microorganism can proliferate, so vacuum packing is often used in combination with other treatment. Vacuum-packing stores food in a vacuum environment, usually in an air-tight bag or bottle. The vacuum environment strips bacteria of oxygen needed for survival, slowing spoiling. Vacuum-packing is commonly used for storing nuts to reduce loss of flavour from oxidation. Curing refers to various food preservation and flavouring processes, especially of meat or fish, by the addition of a combination of salt, sugar, nitrates or nitrite. Many curing processes also involve smoking.

History

Food curing dates back to ancient times, both in the form of smoked meat and as salt-cured meat. Although the ancient people curing the meat did not know this, it was actually nitrates present in the salt that helped the curing process. The Native Americans used to hang their meat at the top of their teepees to increase the amount of smoke coming into contact with the food.

Chemical Actions

Salt

According to the Oklahoma Cooperative Extension Service, salt (sodium chloride; chemical formula: NaCl) is the "primary ingredient used in meat curing". Salt works by dehydrating the meat, thus preventing the growth of bacteria, and it creates an inhospitable osmotic pressure through the cell wall of the bacterium. Doing this requires a concentration of salt of nearly 20%. In addition, salt causes the soluble meat proteins to come to the surface of the meat cut and then solidify, which is what gives sausage its characteristic skin. Finally, salt slows the oxidation process, effectively preventing the meat from going rancid. The sugar added to meat for the purpose of curing it comes in many forms, including honey, corn syrup solids, and maple syrup. However, with the exception of bacon, it does not contribute much to the flavour, but it does alleviate the harsh flavour of the salt. Sugar also contributes to the growth of beneficial bacteria like *Lactobacillus* by feeding them.

Nitrates and Nitrites

Nitrates and nitrites not only help kill bacteria, but also produce a characteristic flavour and give meat a pink or red colour. Nitrate (NO_3^-), generally supplied by sodium nitrate or potassium nitrate, is used as a source for nitrite (NO_2^-). The nitrite further breaks down in the meat into nitric oxide (NO), which then binds to the iron atom in the center of myoglobin's heme group, reducing oxidation and causing a reddish-brown colour (nitrosomyoglobin) when raw, and the characteristic cooked-ham pink colour (nitrosohemochrome or nitrosyl-heme) when cooked. The addition of ascorbate to cured meat reduces formation of nitrosamines, but increases the nitrosylation of iron. The use of nitrates in food preservation is controversial, though, due to the potential for the formation of nitrosamines when the preserved food is cooked at high temperature. The usage of either compound is therefore carefully regulated; for example, in the United States, the concentration of nitrates and nitrites is generally limited to 200 ppm or lower. However, they are considered irreplaceable in the prevention of botulinum poisoning from consumption of cured dry sausages by preventing spore germination.

A 2007 study by Columbia University suggests a link between eating cured meats and chronic obstructive pulmonary disease. Nitrites were posited as a possible cause.

Smoke

Meat can also be preserved by smoking, which is to cook the meat in the presence of smoke. (Thus, the meat is usually near a fire.) Slow-cooking meat while smoking it will also keep it tender. One method of smoking calls for a smokehouse with damp wood chips or sawdust. Salting or curing draws moisture from the meat through a process of osmosis. Meat is cured with salt or sugar, or a combination of the two. Nitrates and nitrites are also often used to cure meat and contribute the characteristic pink colour, as well as inhibition of Clostridium botulinum.

Sugar

Sugar is used to preserve fruits, either in syrup with fruit such as apples, pears, peaches, apricots, plums or in crystallized form where the preserved material is cooked in sugar to the point of crystallisation and the resultant product is then stored dry. This method is used for the skins of citrus fruit (candied peel), angelica and ginger. A modification of this process produces glacé fruit such as glacé cherries where the fruit is preserved in sugar but is then extracted from the syrup and sold, the preservation being maintained by the sugar content of the fruit and the superficial coating of syrup. The use of sugar is often combined with alcohol for preservation of luxury products such as fruit in brandy or other spirits. These should not be confused with fruit flavored spirits such as cherry brandy or Sloe gin.

Artificial Food Additives

Preservative food additives can be *antimicrobial*; which inhibit the growth of bacteria or fungi, including mold, or *antioxidant*; such as oxygen absorbers, which inhibit the oxidation of food constituents.

Common antimicrobial preservatives include calcium propionate, sodium nitrate, sodium nitrite, sulfites (sulphur dioxide, sodium bisulfite, potassium hydrogen sulfite, etc.) and disodium EDTA. Antioxidants include BHA and BHT. Other preservatives include formaldehyde (usually in solution), glutaraldehyde (kills insects), ethanol and methylchloroisothiazolinone.

Pickling

Pickling is a method of preserving food in an edible anti-microbial liquid. Pickling can be broadly categorized as chemical pickling for example, In chemical pickling, the food is placed in an edible liquid that inhibits or kills bacteria and other micro-organisms.

Typical pickling agents include brine (high in salt), vinegar, alcohol, and vegetable oil, especially olive oil but also many other oils. Many chemical pickling processes also involve heating or boiling so that the food being preserved becomes saturated with the pickling agent. Common chemically pickled foods include cucumbers, peppers, corned beef, herring, and eggs, as well mixed vegetables such as piccalilli.

In fermentation pickling, the food itself produces the preservation agent, typically by a process that produces lactic acid. Fermented pickles include sauerkraut, nukazuke, kimchi, surströmming, and curtido. Some pickled cucumbers are also fermented.

In commercial pickles, a preservative like sodium benzoate or EDTA may also be added to enhance shelf life.

Lye

Sodium hydroxide (lye) makes food too alkaline for bacterial growth. Lye will saponify fats in the food, which will change its flavour and texture. Lutefisk uses lye in its preparation, as do some olive recipes. Modern recipes for century eggs also call for lye. Masa harina and hominy use agricultural lime in their preparation and this is often misheard as 'lye'.

Canning and Bottling

Canning involves cooking food, sealing it in sterile cans or jars, and boiling the containers to kill or weaken any remaining bacteria as a form of sterilization. It was invented by Nicolas Appert . Foods have varying degrees of natural protection against spoilage and may require that the final step occur in a pressure cooker. High-acid fruits like strawberries require no preservatives to can and only a short boiling cycle, whereas marginal fruits such as tomatoes require longer boiling and addition of other acidic elements. Low acid foods, such as vegetables and meats require pressure canning. Food preserved by canning or bottling is at immediate risk of spoilage once the can or bottle has been opened.

Lack of quality control in the canning process may allow ingress of water or micro-organisms. Most such failures are rapidly detected as decomposition within the can causes gas production and the can will swell or burst. However, there have been examples of poor manufacture (underprocessing) and poor hygiene allowing contamination of canned food by the obligate anaerobe *Clostridium botulinum*, which produces an acute toxin within the food, leading to severe illness or death. This organism produces no gas or obvious taste and remains undetected by taste or smell. Its toxin is denatured by cooking, though. Cooked mushrooms, handled poorly and then canned, can support the growth of Staphylococcus aureus, which produces a toxin that is not destroyed by canning or subsequent reheating.

Jellying

Food may be preserved by cooking in a material that solidifies to form a gel. Such materials include gelatine, agar, maize flour and arrowroot flour. Some foods naturally form a protein gel when cooked such as eels and elvers, and sipunculid worms which are a delicacy in the town of Xiamen in Fujian province of the People's Republic of China. Jellied eels are a delicacy in the East End of London where they are eaten with mashed potatoes. Potted meats in aspic, (a gel made from gelatine and clarified meat broth) were a common way of serving meat off-cuts in the UK until the 1950s. Many jugged meats are also jellied.

Potting

A traditional British way of preserving meat (particularly shrimp) is by setting it in a pot and sealing it with a layer of fat. Also common is potted chicken liver; compare pâté.

Potted Meat

A potted meat food product or potted meat is a food made using a method of food preservation -canning, consisting of cooked meat product, seasoned, often creamed, minced, or ground, which is filled into cans, sealed and heat processed in a retort to commercial sterility. Various meats such as beef, pork, chicken, turkey and variety (nonskeletal) meats are used. It is produced internationally as a source of affordable meat. Its long shelf life and precooking makes it suitable for emergency food supplies, and for military and camping uses, although the high content of fat, protein, and/or preservatives may make it unsuitable for frequent consumption. The final product typically has a spreadable consistency, and typically contains high amounts of salt, as a preservative.

Reputation

Canned meats have a mixed reputation on account of the taste, texture, ingredients, preparation and nutrition. The canning process produces a product with a generally homogeneous texture and flavour. The low-cost ingredients used also affect the quality. For example, mechanically separated chicken or turkey is a paste-like product made by forcing crushed bone and tissue through a sieve to separate bone from tissue. In the United States, mechanically separated poultry has been used in poultry products since 1969, after the National Academy of Sciences found it safe for use. On November 3, 1995, the Food Safety and Inspection Service (FSIS) of the U.S. Department of Agriculture (USDA) published a final rule in the Federal Register on mechanically separated poultry, stating that it was safe to use without restrictions. However, it must be labelled as "mechanically separated chicken or turkey" in the ingredient statement. The final rule became effective on November 4, 1996.

Ingredients

- Armour Star: Mechanically separated chicken, beef tripe, partially defatted cooked beef fatty tissue, beef hearts, water, partially defatted cooked pork fatty tissue, salt, and less than 2 percent: mustard, natural flavourings, dried garlic, dextrose, sodium erythorbate, and sodium nitrite.
- Hormel: Beef tripe, mechanically separated chicken, beef hearts, partially defatted cooked beef fatty tissue, meat broth, vinegar, salt, flavouring, sugar, and sodium nitrite.
- Libby's: Mechanically separated chicken, pork skin, partially defatted cooked pork fatty tissue, partially defatted cooked beef fatty tissue, vinegar, less than 2% of: salt, spices, sugar, flavourings, sodium erythorbate and sodium nitrite.

Potted Shrimps

Potted shrimps are a traditional Lancastrian dish made with brown shrimp flavoured with mace. The dish consists of brown shrimp in mace-flavoured butter, which has set in a small pot. Cayenne pepper and nutmeg may also be included. It is traditionally eaten with bread. The butter acts as a preservative. Potted shrimp was a favourite dish of Ian Fleming who passed on his predeliction for the delicacy to his famous fictional creation James Bond. Fleming reputedly used to eat the dish at Scotts Restaurant on Mount Street in London where it is still served to this day.

Jugging

In cooking, jugging is the process of stewing meat (usually whole animals, mainly game, or fish) for a long time in a tightly covered container, such as a casserole or an earthenware jug. Sometimes the cooking liquid includes some of the animal's blood. In French, such a stew of a game animal, thickened with the animal's blood, is known as a civet. One common traditional dish that involves jugging is Jugged Hare (known as civet de lièvre in France), which is a whole hare, cut into pieces, marinated and cooked with red wine and juniper berries in a tall jug that stands in a pan of water. It is traditionally served with the hare's blood (or the blood is added right at the very end of the cooking process) and Port wine.

Jugged Hare is described in the influential 18th century cookbook, *The Art of Cookery* by Hannah Glasse, with a recipe titled, "A Jugged Hare," that begins, "Cut it into little pieces, lard them here and there...." The recipe goes on to describe cooking the pieces of hare in water in a jug that is set within a bath of boiling water to cook for three hours. Beginning in the nineteenth century, Glasse has been widely credited with having started the recipe with the words "First, catch your hare," as in this citation. This attribution is apocryphal.

However, having a freshly caught, or shot, hare enables one to obtain its blood. A freshly killed hare is prepared for jugging by removing its entrails and then hanging it in a larder by its hind legs, which causes the blood to accumulate in the chest cavity. One method of preserving the blood after draining it from the hare (since the hare itself is usually hung for a week or more) is to mix it with red wine vinegar in order to prevent it coagulating, and then to store it in a freezer.

Many other British cookbooks from before the middle of the 20th century have recipes for Jugged Hare. Merle and Reitch have this to say about Jugged Hare, for example:

The best part of the hare, when roasted, is the loin and the thick part of the hind leg; the other parts are only fit for stewing, hashing, or jugging. It is usual to roast a hare first, and to stew or jug the portion which is not eaten the first day.

To Jug A Hare. This mode of cooking a hare is very desirable when there is any doubt as to its age, as an old hare, which would be otherwise uneatable, may be made into an agreeable dish.

In 2006, a survey of 2021 people for the television channel UKTV Food found that only 1.6% of the people aged under 25 recognized Jugged Hare by name. 7 out of 10 of those people stated that they would refuse to eat Jugged Hare if it were served at the house of a friend or a relative.

Jugged Rabbit (civet de lapin) is an alternative to Jugged Hare. It is considered a speciality of the cuisine of Martinique.

Another jugged dish, also traditional in the United Kingdom, is Jugged Kippers, which is kippers (with the heads and tails removed) in a covered jug, cooked in boiling water. Recipe books recommend jugging kippers as one way of avoiding the strong smell that kippers have.Meat can be preserved by jugging, the process of stewing the meat (commonly game or fish) in a covered earthenware jug or casserole. The animal to be jugged is usually cut into pieces, placed into a tightly-sealed jug with brine or gravy, and stewed. Red wine and/or the animal's own blood is sometimes added to the cooking liquid. Jugging was a popular method of preserving meat up until the middle of the 20th century.

Irradiation

Irradiation of food is the exposure of food to ionizing radiation; either high-energy electrons or X-rays from accelerators, or by gamma rays (emitted from radioactive sources as Cobalt-60 or Caesium-137). The treatment has a range of effects, including killing bacteria, molds and insect pests, reducing the ripening and spoiling of fruits, and at higher doses inducing sterility. The technology may be compared to pasteurization; it is sometimes called 'cold pasteurization', as the product is not heated. Irradiation is not effective against viruses or prions, it cannot eliminate toxins already formed by microorganisms, and is only useful for food of high initial quality.

The radiation process is unrelated to nuclear energy, but it may use the radiation emitted from radioactive nuclides produced in nuclear reactors. Ionizing radiation is hazardous to life (hence its usefulness in sterilisation); for this reason irradiation facilities have a heavily shielded irradiation room where the process takes place. Radiation safety procedures ensure that neither the workers in such facility nor the environment receive any radiation dose from the facility. Irradiated food does not become radioactive, and national and international expert bodies have declared food irradiation as wholesome. However, the wholesomeness of consuming such food is disputed by opponents and consumer organizations. National and international expert bodies have declared food irradiation as 'wholesome'; UN-organizations as WHO and FAO are endorsing to use food irradiation. International legislation on whether food may be irradiated or not varies worldwide from no regulation to full banning. Irradiation may allow lower quality or contaminated foodstuffs to be rendered marketable.

It is estimated that about 500,000 tons of food items are irradiated per year worldwide in over 40 countries. These are mainly spices and condiments with an increasing segment of fresh fruit irradiated for fruit fly quarantine.

Pulsed Electric Field Processing

Pulsed electric field (PEF) processing is a method for processing cells by means of brief pulses of a strong electric field. PEF holds potential as a type of low temperature alternative pasteurization process for sterilizing food products. In PEF processing, a substance is placed between two electrodes, then the pulsed electric field is applied. The electric field enlarges the pores of the cell membranes which kills the cells and releases their contents. PEF for food processing is a developing technology still being researched. There have been limited industrial applications of PEF processing for the pasteurization of fruit juices.

Modified Atmosphere

Modifying atmosphere is a way to preserve food by operating on the atmosphere around it. Salad crops which are notoriously difficult to preserve are now being packaged in sealed bags with an atmosphere modified to reduce the oxygen (O_2) concentration and increase the carbon dioxide (CO_2) concentration. There is concern that although salad vegetables retain their appearance and texture in such conditions, this method of preservation may not retain nutrients, especially vitamins. Grains may be preserved using carbon dioxide by one of two methods; either using a block of dry ice placed in the bottom and the can is filled with grain or the container can be purged from the bottom by gaseous carbon dioxide from a cylinder or bulk supply vessel.

Carbon dioxide prevents insects, and depending on concentration, mold, and oxidation from damaging the grain. Grain stored in this way can remain edible for five years.

Nitrogen gas (N_2) at concentrations of 98% or higher is also used effectively to kill insects in grain through hypoxia . However, carbon dioxide has an advantage in this respect as it kills organisms through hypercarbia and depending on concentration hypoxia and, requiring concentrations of above 35% , or so. This makes carbon dioxide preferable for fumigation in situations where a hermetic seal cannot be maintained.

Burial in the Ground

Burial of food can preserve it due to a variety of factors: lack of light, lack of oxygen, cool temperatures, pH level, or desiccants in the soil. Burial may be combined with other methods such as salting or fermentation. Many root vegetables are very resistant to spoilage and require no other preservation than storage in cool dark conditions, for example by burial in the ground, such as in a storage clamp.

Century eggs are created by placing eggs in alkaline mud (or other alkaline substance) resulting in their "inorganic" fermentation through raised pH instead of spoiling. The fermentation preserves them and breaks down some of the complex, less flavourful proteins and fats into simpler more flavourful ones.

Most foods can be preserved in soil that is very dry and salty (thus a desiccant), or soil that is frozen.

Cabbage was traditionally buried in the fall in northern farms in the USA for preservation. Some methods keep it crispy while other methods produce sauerkraut. A similar process is used in the traditional production of kimchi. Sometimes meat is buried under conditions which cause preservation. If buried on hot coals or ashes, the heat can kill pathogens, the dry ash can desiccate, and the earth can block oxygen and further contamination. If buried where the earth is very cold, the earth acts like a refrigerator.

Controlled Use of Micro-organism

Some foods, such as many cheeses, wines, and beers will keep for a long time because their production uses specific micro-organisms that combat spoilage from other less benign organisms. These micro-organisms keep pathogens in check by creating an environment toxic for themselves and other micro-organisms by producing acid or alcohol. Starter micro-organisms, salt, hops, controlled (usually cool) temperatures, controlled (usually low) levels of oxygen and/or other methods are used to create the specific controlled conditions that will support the desirable organisms that produce food fit for human consumption.

High Pressure Food Preservation

High pressure food preservation refers to high pressure used for food preservation. "Pressed inside a vessel exerting 70,000 pounds per square inch (480 MPa) or more, food can be processed so that it retains its fresh appearance, flavour, texture and nutrients while disabling harmful microorganisms and slowing spoilage." By 2001, adequate commercial equipment was developed so that by 2005 the process was being used for products ranging from orange juice to guacamole to deli meats and widely sold.

5

Food Contaminants Testing

Food contamination refers to the presence in food of harmful chemicals and microorganisms which can cause consumer illness. This article addresses the chemical contamination of foods, as opposed to microbiological contamination, which can be found under Foodborne illness. A separate issue isgenetically modified food, or the presence in foods of ingredients from genetically modified organisms, also referred to as a form of food contamination.

Impact

The impact of chemical contaminants on consumer health and well-being is often apparent only after many years of prolonged exposure at low levels (e.g. cancer). Chemical contaminants present in foods are often unaffected by thermal processing (unlike most microbiological agents). Chemical contaminants can be classified according to the source of contamination and the mechanism by which they enter the food product.

Agrochemicals

Agrochemicals are chemicals used in agricultural practices and animal husbandry with the intent to increase crops and reduce costs. Such agents includepesticides (e.g. insecticides, herbicides, rodenticides), plant growth regulators, veterinary drugs (e.g. nitrofuran, fluoroquinolones, malachite green, chloramphenicol), and bovine somatotropin (rBST).

Environmental Contaminants

Environmental contaminants are chemicals that are present in the environment in which the food is grown, harvested, transported, stored, packaged, processed, and consumed. The physical contact of the food with its environment results in its contamination. Possible sources of contamination are:

Air: radionuclides (Caesium, Strontium), polycyclic aromatic hydrocarbons (PAH).

Water: arsenic, mercury.

Soil: cadmium, nitrates, perchlorates.

Polychlorinated biphenyls (PCB), dioxins, and polybrominated diphenyl ethers (PBDE) are ubiquitous chemicals, which are present in air, water, soil, and the entire biosphere.

Packaging materials: antimony, tin, lead, perfluorooctanoic acid (PFOA), semicarbazide, benzophenone, isopropylthioxanthone (ITX), bisphenol A.

Processing/cooking equipment: copper, or other metal chips, lubricants, cleaning and sanitizing agents.

Naturally occurring toxins: mycotoxins, phytohaemagglutinin, pyrrolizidine alkaloids, grayanotoxin, mushroom toxins, scombrotoxin (histamine), ciguatera, shellfish toxins (see shellfish poisoning), tetrodotoxin, among many others.

Banned Pesticides and Carcinogens

There are many cases of banned pesticides or carcinogens found in foods.

- Greenpeace exposed in 2006 in China that 25% of surveyed supermarkets agricultural products contained banned pesticides. Over 70% of tomatoes that tested were found to have the banned pesticide Lindane, and almost 40% of the samples had a mix of three or more types of pesticides. Fruits were also tested in this investigation. Tangerines, strawberries and Kyofung grapes samples were found contaminated by banned pesticides, including the highly toxic Methamidophos. These fruits can also be found in Hong Kong market. Greenpeace says there exists no comprehensive monitoring on fruit produce in the Hong Kong as of 2006.
- In India, soft drinks were found contaminated with high levels of pesticides and insecticides, including lindane, DDT, malathion and chlorpyrifos.
- News of Formaldehyde, a carcinogen was found in Vietnamese national dish, Pho, broke in 2007 Vietnam food scare. Vegetables

and fruits were also found to have banned pesticides. "Health agencies have known that Vietnamese soy sauce, the country's second most popular sauce after fish sauce, has been chock full of cancer agents since at least 2001," thundered the Thanh Niendaily. "Why didn't anyone tell us?" The carcinogen in Asian sauces is 3-MCPD and its metabolite 1,3-DCP, which has been an ongoing problem before 2000 affecting multiple continents.

- 2005 Indonesia food scare, carcinogenic formaldehyde was added as a preservative to noodles, tofu, salted fish, and meatballs.
- 2008 Chinese milk scandal.

Hair in Food

There is a heavy stigma attached to the presence of hair in food in most societies. There is a risk that it may induce choking and vomiting, and also that it may be contaminated by toxic substances. Views differ as to the level of risk it poses to the inadvertent consumer.

In most countries, people working in the food industry are required to cover their hair. When people are served food which contains hair in restaurants or cafés, it is usual for them to complain to the staff. Despite this, it is not a valid ground on which to sue the restaurant in the United States but in the United Kingdom it breaks the regulations of the UK Food Safety Act 1990 as it has been known to cause food poisoning and people can sue over this.

There are a range of possible reasons for the objection to hair in food, ranging from cultural taboos to the simple fact that it is difficult to digest and unpleasant to eat. It may also be interpreted as a sign of more widespread problems with hygiene. The introduction of complete capture hairnets is believed to have resulted in a decrease in incidents of contamination of this type.

Sometimes protein from human hair is used as a food ingredient, in bread and other such similar products. Such use of human hair in food is forbidden in Islam. Historically in Judaism, finding hair in food was a sign of bad luck.

Processing Contaminants

Processing contaminants are generated during the processing of foods (e.g. heating, fermentation). They are absent in the raw materials, and are formed by chemical reactions between natural and/or added food constituents during processing. The presence of these contaminants in processed foods cannot be entirely avoided. Technological processes can be adjusted and/or optimized, however, in order to reduce the levels of formation of processing contaminants.

Examples are: nitrosamines, polycyclic aromatic hydrocarbons (PAH), heterocyclic amines, histamine, acrylamide, furan, benzene,trans fat, monochloropropanediol (MCPD), semicarbazide, 4-hydroxynonenal (4-HNE), and ethyl carbamate. There is also the possibility of metal chips from the processing equipment contaminating food. These can be identified using metal detection equipment. In many conveyor lines, the line will be stopped, or when weighing the product with a Check weigher, the item can be rejected for being over-or underweight or because small pieces of metal are detected within it.

Emerging Food Contaminants

While many food contaminants have been known for decades, the formation and presence of certain chemicals in foods has been discovered relatively recently. These are the so-called emerging food contaminants, e.g. acrylamide, furan, benzene, perchlorate, perfluorooctanoic acid (PFOA), 3-monochloropropane-1,3-diol (3-MCPD), 4-hydroxynonenal and (4-HNE).

Safety and Regulation

Acceptable Daily Intake (ADI) levels and tolerable concentrations of contaminants in individual foods are determined on the basis of the "No Observed Adverse Effect Level" (NOAEL) in animal experiments, by using a safety factor (usually 100). The maximum concentrations of contaminants allowed by legislation are often well below toxicological tolerance levels, because such levels can often be reasonably achieved by using good agricultural and manufacturing practices.

The establishment of ADIs for certain emerging food contaminants is currently an active area of research and regulatory debate.

Food Safety

Food safety is a scientific discipline describing handling, preparation, and storage of food in ways that prevent foodborne illness. This includes a number of routines that should be followed to avoid potentially severe health hazards. Food can transmit disease from person to person as well as serve as a growth medium for bacteria that can cause food poisoning. Debates on genetic food safety include such issues as impact of genetically modified food on health of further generations and genetic pollution of environment, which can destroy natural biological diversity. In developed countries there are intricate standards for food preparation, whereas in lesser developed countries the main issue is simply the availability of adequate safe water, which is usually a critical item. In theory food poisoning is 100% preventable.

Key Principles

Five Key Principles

The five key principles of food hygiene, according to WHO, are:

1. Prevent contaminating food with pathogens spreading from people, pets, and pests.
2. Separate raw and cooked foods to prevent contaminating the cooked foods.
3. Cook foods for the appropriate length of time and at the appropriate temperature to kill pathogens.
4. Store food at the proper temperature.
5. Use safe water and raw materials

ISO 22000

ISO 22000 is a standard developed by the International Organization for Standardization dealing with food safety. This is a general derivative of ISO 9000. ISO 22000 standard: The ISO 22000 international standard specifies the requirements for a food safety management system that involves interactive communication, system management, prerequisite programs, HACCPprinciples.

Incidence

A 2003 World Health Organization (WHO) report concluded that about 40% of reported food poisoning outbreaks in the WHO European Region occur in private homes. According to the WHO and CDC, in the USA alone, annually, there are 76 million cases of foodborne illness leading to 325,000 hospitalizations and 5,000 deaths.

Regulatory Agencies

Australia

Australian Food Authority is working toward ensuring that all food businesses implement food safety systems to ensure food is safe to consume in a bid to halt the increasing incidence of food poisoning, this includes basic food safety training for at least one person in each business. Smart business operators know that basic food safety training improves the bottom line, staff take more pride in their work; there is less waste; and customers can have more confidence in the food they consume.

Food Safety training in units of competence from a relevant training package, must be delivered by a Registered Training Organization (RTO) to enable staff to be issued with a nationally-recognised unit of competency code on their certificate. Generally this training can be completed in less than one day. Training options are available to suit the needs of everyone. Training may be carried out in-house for a group, in a public class, via correspondence or online. Basic food safety training includes:

- Understanding the hazards associated with the main types of food and the conditions to prevent the growth of bacteria which can cause food poisoning
- The problems associated with product packaging such as leaks in vacuum packs, damage to packaging or pest infestation, as well as problems and diseases spread by pests.
- Safe food handling. This includes safe procedures for each process such as receiving, re-packing, food storage, preparation and cooking, cooling and re-heating, displaying products, handling products when serving customers, packaging, cleaning and sanitizing, pest control, transport and delivery. Also the causes of cross contamination.
- Catering for customers who are particularly at risk of food-borne illness, including allergies and intolerance.
- Correct cleaning and sanitizing procedures, cleaning products and their correct use, and the storage of cleaning items such as brushes, mops and cloths.
- Personal hygiene, hand washing, illness, and protective clothing.

People responsible for serving unsafe food can be liable for heavy fines under this new legislation, consumers are pleased that industry will be forced to take food safety seriously.

China

Food safety is a growing concern in Chinese agriculture. The Chinese government oversees agricultural production as well as the manufacture of food packaging, containers, chemical additives, drug production, and business regulation. In recent years, the Chinese government attempted to consolidate food regulation with the creation of the State Food and Drug Administration in 2003, and officials have also been under increasing public and international pressure to solve food safety problems. However, it appears that regulations are not well known by the trade. Labels used for "green" food, "organic" food and "pollution-free" food are not well recognized by traders and many are unclear about their meaning. A survey by the World Bank found that supermarket managers had difficulty in obtaining produce that met safety requirements and found that a high percentage of produce did not comply with established standards.

Traditional marketing systems, whether in China or the rest of Asia, presently provide little motivation or incentive for individual farmers to make improvements to either quality or safety as their produce tends to get grouped together with standard products as it progresses through the marketing channel. Direct linkages between farmer groups and traders or ultimate buyers, such as supermarkets, can help avoid this problem.

Governments need to improve the condition of many markets through upgrading management and reinvesting market fees in physical infrastructure. Wholesale markets need to investigate the feasibility of developing separate sections to handle fruits and vegetables that meet defined safety and quality standards.

European Union

The parliament of the European Union (EU) makes legislation in the form of directives and regulations, many of which are mandatory for member states and which therefore must be incorporated into individual countries' national legislation. As a very large organisation that exists to remove barriers to trade between member states, and into which individual member states have only a proportional influence, the outcome is often seen as an excessively bureaucratic 'one size fits all' approach. However, in relation to food safety the tendency to err on the side of maximum protection for the consumer may be seen as a positive benefit. The EU parliament is informed on food safety matters by the European Food Safety Authority.

Individual member states may also have other legislation and controls in respect of food safety, provided that they do not prevent trade with other states, and can differ considerably in their internal structures and approaches to the regulatory control of food safety.

Germany

The Federal Ministry of Food, Agriculture and Consumer Protection (BMELV) is a Federal Ministry of the Federal Republic of Germany. History: Founded as Federal Ministry of Food, Agriculture and Foresting in 1949, this name did not change until 2001. Then the name changed to Federal Ministry of Consumer Protection, Food and Agriculture. At the 22nd of November 2005, the name got changed again to its current state: Federal Ministry of Food, Agriculture and Consumer Protection.

The reason for this last change was that all the resorts should get equal ranking which was achieved by sorting the resorts alphabetically. Vision: A balanced and healthy diet with safe food, distinct consumer rights and consumer information for various areas of life, and a strong and sustainable agriculture as well as perspectives for our rural areas are important goals of the Federal Ministry of Food, Agriculture and Consumer Protection (BMELV). The Federal Office of Consumer Protection and Food Safety is under the control of the Federal Ministry of Food, Agriculture and Consumer Protection. It exercises several duties, with which it contributes to safer food and thereby intensifies health-based consumer protection in Germany.

Food can be manufactured and sold within Germany without a special permission, as long as it does not cause any damage on consumers' health and meets the general standards set by the legislation. However, manufacturers, carriers, importers and retailers are responsible for the food they pass into circulation. They are obliged to ensure and document the safety and quality of their food with the use of in-house control mechanisms.

Hong Kong

In Hong Kong SAR, the Centre for Food Safety is in charge of ensuring food sold is safe and fit for consumption.

Pakistan

Pakistan does not have an integrated legal framework but has a set of laws, which deals with various aspects of food safety. These laws, despite the fact that they were enacted long time ago, have tremendous capacity to achieve at least minimum level of food safety. However, like many other laws, these laws remain very poorly enforced. There are four laws that specifically deal with food safety. Three of these laws directly focus issues related to food safety, while the fourth one namely Pakistan Standards and Quality Control Authority Act, is indirectly relevant to food safety. The Pure Food Ordinance, 1960 consolidates and amends the law in relation to the preparation and the sale of foods. All provinces and some northern areas have adopted this law with certain amendments. Its aim is to ensure purity of food being supplied to people in the market and, therefore, provides for preventing adulteration. The Pure Food Ordinance 1960 does not apply to cantonment areas.

There is separate law for cantonments called "The Cantonment Pure Food Act, 1966". There is no substantial difference between the Pure Food Ordinance 1960 and The Cantonment Pure Food Act. Even the rules of operation are very much similar. Pakistan Hotels and Restaurant Act, 1976 applies to all hotels and restaurants in Pakistan and seeks to control and regulate the rates and standard of service(s) by hotels and restaurants.

In addition to other provisions, under section 22(2), the sale of food or beverages that are contaminated, not prepared hygienically or served in utensils that are not hygienic or clean is an offence. There are no express provisions for consumer complaints in the Pakistan Restaurants Act, 1976, Pakistan Penal Code, 1860 and Pakistan Standards and Quality Control Authority Act, 1996. However, the laws do not prevent citizens from lodging complaints with the concerned government officials. However, the consideration and handling of complaints is a matter of discretion of the officials.

South Korea

Korea Food and Drug Administration: Korea Food & Drug Administration (KFDA) is working for food safety since 1945. It is part of the Government of South Korea.

IOAS-Organic Certification Bodies Registered in KFDA: "Organic" or related claims can be labelled on food products when organic certificates are considered as valid by KFDA. KFDA admits organic certificates which can be issued by 1) IFOAM (International Federation of Organic Agriculture Movement) accredited certification bodies 2) Government accredited certification bodies-328 bodies in 29 countries have been registered in KFDA.

Food Import Report: According to Food Import Report, it is supposed to report or register what you import. Competent authority is as followed:

- Imported Agricultural Products, Processed Foods, Food Additives, Utensils, Containers & Packages or Health Functional Foods
  — KFDA (Korea Food and Drug Administration)
- Imported Livestock, Livestock products (including Dairy products)

— NVRQS (National Veterinary Research and Quarantine Service)-Packaged meat, milk & dairy products (butter, cheese), hamburger patties, meat ball and other processed products which are stipulated by Livestock Sanitation Management Act

- Imported Marine products

— NFIS (National Fisheries Products Quality Inspection Service)-Fresh, chilled, frozen, salted, dehydrated, eviscerated marine produce which can be recognized its characteristics.

National Institute of Food and Drug Safety Evaluation

National Institute of Food and Drug Safety Evaluation (NIFDS) is functioning as well. The National Institute of Food and Drug Safety Evaluation is a national organization for toxicological tests and research. Under the Korea Food & Drug Administration, the Institute performs research on toxicology, pharmacology, and risk analysis of foods, drugs, and their additives.

The Institute strives primarily to understand important biological triggering mechanisms and improve assessment methods of human exposure, sensitivities, and risk by

(1) conducting basic, applied, and policy research that closely examines biologically triggering harmful effects on the regulated products such as foods, food additives, and drugs, and

(2) operating the national toxicology program for the toxicological test development and inspection of hazardous chemical ubstances assessments.

The Institute ensures safety by

(1) investigation and research on safety by its own researchers,

(2) contract research by external academicians and research centres.

United States

The US food system is regulated by numerous federal, state and local officials. Although the US food safety system is one of the best in the world, it is lacking in "organization, regulatory tools, and resources to address food borne illness."

Federal Level Regulation

The Food and Drug Administration publishes the Food Code, a model set of guidelines and procedures that assists food control jurisdictions by providing a scientifically sound technical and legal basis for regulating the retail and food service industries, including restaurants, grocery stores and institutional food service providers such as nursing homes.

Regulatory agencies at all levels of government in the United States use the FDA Food Code to develop or update food safety rules in their jurisdictions that are consistent with national food regulatory policy. According to the FDA, 48 of 56 states and territories, representing 79% of the U.S. population, have adopted food codes patterned after one of the five versions of the Food Code, beginning with the 1993 edition.

In the United States, federal regulations governing food safety are fragmented and complicated, according to a February 2007 report from the Government Accountability Office. There are 15 agencies sharing oversight responsibilities in the food safety system, although the two primary agencies are the U.S. Department of Agriculture (USDA) Food Safety and Inspection Service (FSIS), which is responsible for the safety of meat, poultry, and processed egg products, and the Food and Drug Administration (FDA), which is responsible for virtually all other foods.

The Food Safety and Inspection Service has approximately 7,800 inspection program personnel working in nearly 6,200 federally inspected meat, poultry and processed egg establishments. FSIS is charged with administering and enforcing the Federal Meat Inspection Act, the Poultry Products Inspection Act, the Egg Products Inspection Act, portions of the Agricultural Marketing Act, the Humane Slaughter Act, and the regulations that implement these laws. FSIS inspection program personnel inspect every animal before slaughter, and each carcass after slaughter to ensure public health requirements are met. In fiscal year (FY) 2008, this included about 50 billion pounds of livestock carcasses, about 59 billion pounds of poultry carcasses, and about 4.3 billion pounds of processed egg products. At U.S. borders, they also inspected 3.3 billion pounds of imported meat and poultry products.

Industry Pressure

There have been concerns over the efficacy of safety practices and food industry pressure on U.S. regulators. A study reported by Reuters found that "the food industry is jeopardizing U.S. public health by withholding information from food safety investigators or pressuring regulators to withdraw or alter policy designed to protect consumers".

A survey found that 25% of U.S. government inspectors and scientists surveyed have experienced during the past year corporate interests forcing their food safety agency to withdraw or to modify agency policy or action that protects consumers. Scientists have observed that management undercuts field inspectors who stand up for food safety against industry pressure. According to Dr. Dean Wyatt, a USDA veterinarian who oversees federal slaughter house inspectors, "Upper level management does not adequately support field inspectors and the actions they take to protect the food supply. Not only is there lack of support, but there's outright obstruction, retaliation and abuse of power."

State and Local Regulation

A number of U.S. states have their own meat inspection programs that substitute for USDA inspection for meats that are sold only in-state. Certain state programs have been criticized for undue leniency to bad practices. However, other state food safety programs supplement, rather than replace, Federal inspections, generally with the goal of increasing consumer confidence in the state's produce. For example, state health departments have a role in investigating outbreaks of food-borne disease bacteria, as in the case of the 2006 outbreak of *Escherichia coli* O157:H7 (bad *E. coli* bacteria) from processed spinach. Health departments also promote better food processing practices to eliminate these threats.

In addition to the US Food and Drug Administration, several states that are major producers of fresh fruits and vegetables (including California, Arizona and Florida) have their own state programs to test produce for pesticide residues.

Restaurants and other retail food establishments fall under state law and are regulated by state or local health departments. Typically these regulations require official inspections of specific design features, best food-handling practices, and certification of food handlers. In some places a letter grade or numerical score must be prominently posted following each inspection. In some localities, inspection deficiencies and remedial action are posted on the Internet.

Manufacturing Control

HACCP Guidelines

The UK Food Standards Agency publishes recommendations as part of its Hazard Analysis and Critical Control Points (HACCP) programme. The relevant guidelines state that:

"Cooking food until the CORE TEMPERATURE is 75 °C or above will ensure that harmful bacteria are destroyed.

However, lower cooking temperatures are acceptable provided that the CORE TEMPERATURE is maintained for a specified period of time as follows:

- 60 °C for a minimum of 45 minutes
- 65 °C for a minimum of 10 minutes
- 70 °C for a minimum of 2 minutes".

Previous guidance from a leaflet produced by the UK Department Of Health "Handling Cooked Meats Safely A Ten Point Plan" also allowed for:

- "75 °C for a minimum of 30 seconds
- 80 °C for a minimum of 6 seconds".

Note that recommended cooking conditions are only appropriate if initial bacterial numbers in the uncooked food are small. Cooking does not overcome poor hygiene.

Consumer Labelling

United Kingdom

Food stuffs in the UK have one of two labels to indicate the nature of the deterioration of the product and any subsequent health issues. EHO Food Hygiene certification is required to prepare and distribute food.

While there is no specified expiry date of such a qualification the changes in legislation it is suggested to update every five years.

Best before indicates a future date beyond which the food product *may* lose quality in terms of taste or texture amongst others, but does not imply any serious health problems if food is consumed beyond this date (within reasonable limits).

Use by indicates a legal date beyond which it is not permissible to sell a food product (usually one that deteriorates fairly rapidly after production) due to the potential serious nature of consumption of pathogens.

Leeway is sometimes provided by producers in stating display until dates so that products are not at their limit of safe consumption on the actual date stated (this latter is voluntary and not subject to regulatory control). This allows for the variability in production, storage and display methods.

United States

With the exception of infant formula and baby foods which must be withdrawn by their expiration date, Federal law does not require expiration dates. For all other foods, except dairy products in some states, freshness dating is strictly voluntary on the part of manufacturers. In response to consumer demand, perishable foods are typically labeled with a Sell by date. It is up to the consumer to decide how long after the Sell by date a package is usable. Other common dating statements are Best if used by, Use-by date, Expiration date, Guaranteed fresh <date>, and Pack date.

Australia and New Zealand

Guide to Food Labelling and Other Information Requirements: This guide provides background information on the general labelling requirements in the Code. The information in this guide applies both to food for retail sale and to food for catering purposes.

Foods for catering purposes means those foods for use in restaurants, canteens, schools, caterers or self-catering institutions, where food is offered for immediate consumption. Labelling and information requirements in the new Code apply both to food sold or prepared for sale in Australia and New Zealand and food imported into Australia and New Zealand. Warning and Advisory Declarations, Ingredient Labelling, Date Marking, Nutrition Information Requirements, Legibility Requirements for Food Labels, Percentage Labelling, Information Requirements for Foods Exempt from Bearing a Label.

How to Enhance Food Safety

There is a number of ways to enhance sell by and use by dates. These include better education of consumers on how to use, transport, and store fresh food products, but also by enhancing the use by and sell by dates by adding to the package smart indicators such as TTIs (Time Temperature Indicators). These show through a visible colour change whether the product is still fresh. TTIs are already in use by retailers and food producers in France (Monoprix and Carrefour), Switzerland (Kneuss), and other countries in western Europe.

Consumer Tips: How to Keep Food Safe

1. One of the simplest measures that any person can take to prevent the spread of foodborne illness is to properly wash his or her hands before preparing or eating any meal. Many people

who believe they are adequately washing their hands are sorely mistaken. According to a study conducted by the American Society of Microbiology, 97% of females and 92% of males said they washed their hands, but those numbers turned out to be 75% of females and 58% of males upon observation. A proper hand-washing technique suggested by the Missouri Department of Health and Senior Services includes using soap and warm water; washing vigorously for 10-20 seconds, making sure to reach all surfaces of the hands including the wrists, between the fingers, and under the finger nails; rinsing well; drying hands with a paper towel; and using a paper towel to turn off the water. It is certainly important to wash your hands before preparing a meal and eating, but hands should also be washed after using the restroom, coughing/sneezing, touching cuts or skin infections, handling raw meat, and touching pets or other animals. Almost half the cases of foodborne illnesses could be prevented by better hand-washing by food handlers. In addition, hand sanitizer is a helpful follow-up to hand-washing, but it should never be used to replace this valuable technique.

2. Some foods should simply never be ingested in the first place because they have such a high risk of containing harmful bacteria that can make people ill. These foods include raw milk, unpasteurized dairy products, unpasteurized juices (such as fresh apple cider), raw meat, and raw cookie dough.
3. In cases of suspected food spoilage, food should never simply be tasted, smelled, or eye-balled in order to determine its safety. It is true that in some cases, the presence of mold or other growths may indicate that a food has reached its expiration date and should not be consumed. However, some foods may look and smell fine, but if they have been stored at room temperature for over 2 hours, microscopic bacteria may have been allowed to grow. The safe temperature for foods in refrigerators is between 35°F and 45°F, and freezers should be kept at 0° F or below.
4. It is not safe to let meats thaw on the counter all day, because this allows any germs present on the food to thrive. Safer alternatives to this practice include thawing the food under running water (70°F or below) for less than 2 hours, placing the food in the refrigerator to thaw, or thawing the food in the microwave as part of the cooking process.

5. It is also important to make sure that raw meats are cooked to the appropriate internal temperature before they are consumed. Safe internal temperatures for various meats include 165°F for poultry, 155°F for ground meat, and 145°F for fish and pork. It is not sufficient to judge the doneness of meat by its internal colour. According to a study performed by the USDA, "25% of hamburgers with a brown internal colour were not cooked to the proper temperature." Instead of simply eye balling the meat, it is essential to use a meat thermometer to judge the safety of consuming the food.
6. The cooking process is often the time that foodborne pathogens are allowed to enter the food we eat because of the prevalence of cross contamination. Cross contamination occurs when a person handling raw meats, eggs, fish, or other foods containing harmful pathogens touches cooking utensils, cutting boards, or cooking surfaces and spreads the pathogens to ready-to-eat foods in the process. This mode of transmission can be interrupted by washing hands after handling raw foods, washing utensils and cutting boards that have come in contact with raw foods, and disinfecting counter surfaces frequently.
7. Leftovers are ideally stored in the refrigerator in shallow containers (2 inches tall or less) so that the cooling process can be accelerated and the buildup of harmful bacteria can be prevented. Storing foods in larger containers may keep foods warm and allow harmful bacteria to grow.

Food Preservation

Food preservation is the process of treating and handling food to stop or slow down spoilage (loss of quality, edibility or nutritional value).

Preservation usually involves preventing the growth of bacteria, yeasts, fungi, and other micro-organisms (although some methods work by introducing benignbacteria, or fungi to the food), as well as retarding the oxidation of fats which cause rancidity. Black preservation also includes processes which inhibit natural discoloration that can occur during food preparation, such as the enzymatic browning reaction in apples after they are cut.

Many processes designed to preserve food will involve a number of food preservation methods. Preserving fruit, by turning it into jam, for example, involves boiling (to reduce the fruit's moisture content and to kill bacteria, yeasts, etc.), sugaring (to prevent their re-growth) and sealing within an airtight jar (to prevent recontamination). There are many traditional methods of preserving food that limit the energy inputs and reduce carbon footprint.

Maintaining or creating nutritional value, texture and flavour is an important aspect of food preservation, although, historically, some methods drastically altered the character of the food being preserved. In many cases these changes have now come to be seen as desirable qualities – cheese, yoghurt and pickled onions being common examples.

Food Fortification

Food fortification is the public health policy of adding micronutrients (essential trace elements and vitamins) to foodstuffs to ensure that minimum dietary requirements are met.

Simple diets based on staple foods with little variation are often deficient in certain nutrients, either because they are not present in sufficient amounts in the soil of a region, or because of the inherent inadequacy of the diet. Addition of micronutrients to staples and condiments can prevent large-scale deficiency diseases in these cases.

Rationale

Several ranges of food supplements are recognised:

- additives which repair a deficit to "normal" levels
- additives which appear to enhance a food
- supplements taken in addition to the normal diet

Many physicians today disagree with the premise that foodstuffs need supplementation, but accept that-for example-added calcium may provide benefit, or that adding folic acid may correct a nutritional deficiency especially in pregnant women.

On a more controversial level, but well founded in scientific basis, is the science of using foods and food supplements to achieve a defined health goal. A common example of this use of food supplements is the extent to which body builders will use amino acid mixtures, vitamins and phytochemicals to enhance natural hormone production, increase muscle and reduce fat.

Moving on from this reasonably accepted usage, there is increasing evidence for the use of food supplements in established medical conditions. This nutritional supplementation using foods as medicine (nutraceuticals) has been effectively used in treating disorders affecting the immune system up to and including cancers. This goes beyond the definition of "food supplement", but should be included for the sake of completeness.

Food Supplements

There are several main groups of food supplements which can be considered:

- Vitamins and co-vitamins
- Essential minerals
- Essential fatty acids
- Essential amino acids
- Glyconutrients
- Phytonutrients
- Enzymes.

Examples of Fortified Foods

Iodised salt has been used in the United States since before World War II.

Folic acid is added to flour in many industrialized countries, and has prevented a significant number of neural tube defects in infants. It is, however, not uniform in its application, with more intake of folic acid through fortified flour among those who were already receiving high amounts through their diet.

Niacin has been added to bread in the USA since 1938 (when voluntary addition started), a programme which substantially reduced the incidence of pellagra. Vitamin D is added to a few foods (especially margarine). Fluoride salts are added to water and toothpastes to prevent tooth decay. Water fluoridation is a controversial topic in some segments of the general public, although less so amongst established scientific bodies.

Calcium is frequently added to fruit juices, carbonated beverages and rice. "Golden rice" is a variety of rice which has been genetically modified to produce beta carotene.

Dietary Supplement

"Food supplement" redirects here. For food additions that alter the flavour, colour or longevity of food, see Food additive.

A dietary supplement, also known as food supplement or nutritional supplement, is a preparation intended to supplement the diet and provide nutrients, such as vitamins, minerals, fibre, fatty acids, or amino acids, that may be missing or may not be consumed in sufficient quantities in a person's diet.

Some countries define dietary supplements as foods, while in others they are defined as drugs or natural health products.

Supplements containing vitamins or dietary minerals are included as a category of food in the Codex Alimentarius, a collection of internationally recognized standards, codes of practice, guidelines and other recommendations relating to foods, food production and food safety.

These texts are drawn up by the Codex Alimentarius Commission, an organization that is sponsored by the Food and Agriculture Organization of the United Nations (FAO) and the World Health Organization (WHO).

6

Food Fermentation Process

Wine is Fermented Grape Juice

Wine can be made from grapes, fruits, berries etc. Most wine, though, is made from grapes. And no matter what the wine is made from, there must be fermentation, that is, that sugar be transformed into alcohol. If the amount of alcohol is relatively low, the result is wine. If it is high, the result is a "distilled liquor," something like gin or vodka. There are red wines, pink wines (also known as "rose" or some-times "blush") and white wines.

Red wine result when the crushed grape skin pulp and seeds of purple or red varieties are allowed to remain with juice during fermentation periods. Pink/rose wine can be produced by removing the non-juice pumace from the must during fermentation. White wines can be made from pigmented grapes by removal of skins, pulp and seeds before juice fermentation. Wines might be "fortified," "sparkling," or "table."

In fortified wines, brandy is added to make the alcohol content higher (around 14 to 30 percent). These are less perishable and may be stable without pasteurization. Wines are termed still or sparkling depending upon the amount of CO_2 they contain. The carbon dioxide may be formed naturally during fermentation or may be added artificially.

Both table and sparkling wines tend to have alcohol contents between 7 and 14 percent. Sparkling wines are the ones with bubbles (greater CO_2), like Champagne. Table wine (which can also be called "still") are the most "natural". The alcohol concentration itself is not sufficient to preserve natural wines, they are pasteurized. The term light wine is also used to describe wine having alcohol content from 5- 10 %.

How is Wine Made?

Growing Grapes

Grapes grow on vines. There are many different types of grapes, but the best wine grape is the European Vitis vinifera. It is considered optimal because it has the right balance of sugar and acid to create a good fermented wine without the addition of sugar or water.

Harvest

Weather is a major factor is determining whether a year is going to be a "good vintage" (or "year"). For example, was there enough heat during the growing season to lead to enough sugar? At harvest time, the short-term effects of weather are quite important. To produce great wine, the fruit should have a high (but not overly high) sugar content ("brix"). Think of raisins. As the fruit dries, the water evaporates. What is left is the sugary fruit. If it rains just at the point the wine grapes are ready, and before the grapes can be harvested, the additional water will cause the water level to increase, and the brix will go down. Not good. (You might ask, why not just add some sugar in the wine making process? Some do. Also considered "not good.")

Every year the wine grape grower plays a game of chance and must decide when to harvest. Simplistically, if you knew it wasn't going to rain, you would just test the brix until it was just right, then harvest. If you harvest too soon, you will probably end up getting a wine too low in alcohol content (there won't have been enough sugar to convert to alcohol). These wines will be "thin." If you delay harvest, there may be too much sugar, which leads to too low acid content. This also affects the taste (and the aging possibilities) of the wine.

Initial Processing of the Grape Juice

Grapes can (and might still) be crushed by stomping on them with your feet in a big vat. But a more practical way is to use a machine which does the job (and at the same time, removes the stems).

What you get may or may not get immediately separated. Skin and seeds might immediately be removed from the juice. Separation may not immediately occur (especially for red wines), since skins and stems are an important source of "tannins" which affect wine's taste and maturity through aging. The skins also determine the colour of the wine.

Maceration (the time spent while skins and seeds are left with the juice) will go on for a few hours or a few weeks. Pressing will then occur. One way to press the grapes is to use a "bladder press," a large cylindrical container that contains bags that are inflated and deflated several times, each time gently squeezing the grapes until all the juice has run free, leaving behind the rest of the grapes. You can also separate solids from juice through the use of a centrifuge.

Crushing and Primary Fermentation

Crushing is the process of gently squeezing the berries and breaking the skins to start to liberate the contents of the berries. Desteming is the process of removing the grapes from the rachis (the stem which holds the grapes). In traditional and smaller-scale wine making, the harvested grapes are sometimes crushed by trampling them barefoot or by the use of inexpensive small scale crushers. These can also destem at the same time. However, in larger wineries, a mechanical crusher/destemmer is used. The decision about desteming is different for red and white wine making. Generally when making white wine the fruit is only crushed, the stems are then placed in the press with the berries.

The presence of stems in the mix facilitates pressing by allowing juice to flow past flattened skins. These accumulate at the edge of the press. For red winemaking, stems of the grapes are usually removed before fermentation since the stems have a relatively high tannin content; in addition to tannin they can also give the wine a vegetal aroma (due to extraction of 2-methoxy-3-isopropylpyrazine which has an aroma reminiscent of green bell peppers.) On occasion, the winemaker may decide to leave them in if the grapes themselves contain less tannin than desired. This is more acceptable if the stems have 'ripened' and started to turn brown. If increased skin extraction is desired, a winemaker might choose to crush the grapes after destemming. Removal of stems first means no stem tannin can be extracted. In these cases the grapes pass between two rollers which squeeze the grapes enough to separate the skin and pulp, but not so much as to cause excessive shearing or tearing of the skin tissues. In some cases, notably with "delicate" red varietals such as Pinot noir or Syrah, all or part of the grapes might be left uncrushed (called "whole berry") to encourage the retention of fruity aromas through partial carbonic maceration.

Most red wines derive their colour from grape skins (the exception being varieties or hybrids of non-vinifera vines which contain juice pigmented with the dark Malvidin 3,5-diglucoside anthocyanin) and therefore contact between the juice and skins is essential for colour extraction. Red wines are produced by destemming and crushing the grapes into a tank and leaving the skins in contact with the juice throughout the fermentation (maceration). It is possible to produce white (colourless) wines from red grapes by the fastidious pressing of uncrushed fruit. This minimizes contact between grape juice and skins (as in the making of *Blanc de noirs* sparkling wine, which is derived from Pinot noir, a red vinifera grape.)

Most white wines are processed without destemming or crushing and are transferred from picking bins directly to the press. This is to avoid any extraction of tannin from either the skins or grapeseeds, as well as maintaining proper juice flow through a matrix of grape clusters rather than loose berries. In some circumstances winemakers choose to crush white grapes for a short period of skin contact, usually for three to 24 hours. This serves to extract flavour and tannin from the skins (the tannin being extracted to encourage protein precipitation without excessive Bentonite addition) as well as Potassium ions, which participate in bitartrate precipitation (cream of tartar). It also results in an increase in the pH of the juice which may be desirable for overly acidic grapes. This was a practice more common in the 1970s than today, though still practiced by some Sauvignon blanc and Chardonnay producers in California.

In the case of rose wines, the fruit is crushed and the dark skins are left in contact with the juice just long enough to extract the colour that the winemaker desires. The must is then pressed, and fermentation continues as if the wine maker was making a white wine.

Yeast is normally already present on the grapes, often visible as a powdery appearance of the grapes. The fermentation can be done with this natural yeast, but since this can give unpredictable results depending on the exact types of yeast that are present, cultured yeast is often added to the must. One of the main problems with the use of wild ferments is the failure for the fermentation to go to completion, that is some sugar remains unfermented. This can make the wine sweet when a dry wine is desired. Frequently wild ferments lead to the production of unpleasant acetic acid (vinegar) production as a by product. During the primary fermentation, the yeast cells feed on the sugars in the must and multiply, producing carbon dioxide gas and alcohol. The temperature during the fermentation affects both the taste of the end product, as well as the speed of the fermentation.

For red wines, the temperature is typically 22 to 25 °C, and for white wines 15 to 18 °C. For every gram of sugar that is converted, about half a gram of alcohol is produced, so to achieve a 12% alcohol concentration, the must should contain about 24% sugars. The sugar percentage of the must is calculated from the measured density, the must weight, with the help of a specialized type of hydrometer called a saccharometer. If the sugar content of the grapes is too low to obtain the desired alcohol percentage, sugar can be added (chaptalization). In commercial winemaking, chaptalization is subject to local regulations.

During or after the alcoholic fermentation, malolactic fermentation can also take place, during which specific strains of bacteria convert malic acid into the milder lactic acid. This fermentation is often initiated by inoculation with desired bacteria.

Pressing

Pressing is the act of applying pressure to grapes or pomace in order to separate juice or wine from grapes and grape skins. Pressing is not always a necessary act in winemaking; if grapes are crushed there is a considerable amount of juice immediately liberated (called free-run juice) that can be used for vinification. Typically this free-run juice is of a higher quality than the press juice. However, most wineries do use presses in order to increase their production (gallons) per ton, as pressed juice can represent between 15%-30% of the total juice volume from the grape.

Presses act by positioning the grape skins or whole grape clusters between a rigid surface and a moveable surface and slowly decrease the volume between the two surfaces. Modern presses dictate the duration and pressure at each press cycle, usually ramping from 0 Bar to 2.0 Bar. Sometimes winemakers choose pressures which separate the streams of pressed juice, called making "press cuts."

As the pressure increases the amount of tannin extracted from the skins into the juice increases, often rendering the pressed juice excessively tannic or harsh. Because of the location of grape juice constituents in the berry (water and acid are found primarily in the mesocarp or pulp, whereas tannins are found primarily in the pericarp, or skin, and seeds), pressed juice or wine tends to be lower in acidity with a higher pH than the free-run juice.

Before the advent of modern winemaking, most presses were basket presses made of wood and operated manually. Basket presses are composed of a cylinder of wooden slats on top of a fixed plate, with a moveable plate that can be forced downward (usually by a central ratcheting threaded screw.) The press operator would load the grapes or pomace into the wooden cylinder, put the top plate in place and lower it until juice flowed from the wooden slats. As the juice flow decreased, the plate was ratcheted down again. This process continued until the press operator determined that the quality of the pressed juice or wine was below standard, or all liquids had been pressed. Since the early 1990s, modern mechanical basket presses have been revived through higher-end producers seeking to replicate the gentle pressing of the historical basket presses. Because basket presses have a relatively compact design, the press cake offers a relatively longer pathway for the juice to travel before leaving the press. It is believed by advocates of basket presses that this relatively long pathway through the grape or pomace cake serves as a filter to solids that would otherwise affect the quality of the press juice. With red wines, the must is pressed after primary fermentation, which separates the skins and other solid matter from the liquid. With white wine, the liquid is separated from the must before fermentation. With rose, the skins may be kept in contact for a shorter period to give colour to the wine, in that case the must may be pressed as well. After a period in which the wine stands or ages, the wine is separated from the dead yeast and any solids that remained (called lees), and transferred to a new container where any additional fermentation may take place.

Pigeage

Pigeage is a French winemaking term for the traditional stomping of grapes in open fermentation tanks. To make certain types of wine, grapes are put through a crusher and then poured into open fermentation tanks. Once fermentation begins, the grape skins are pushed to the surface by carbon dioxide gases released in the fermentation process. This layer of skins and other solids is known as the cap. As the skins are the source of the tannins, the cap needs to be mixed through the liquid each day, or "punched," which traditionally is done by stomping through the vat.

Cold and Heat Stabilization

Cold stabilization is a process used in winemaking to reduce tartrate crystals (generally potassium bitartrate) in wine. These tartrate crystals look like grains of clear sand, and are also known as "wine crystals" or "wine diamonds". They are formed by the union of tartaric acid and potassium, and may appear to be sediment in the wine, though they are not. During the cold stabilizing process after fermentation, the temperature of the wine is dropped to close to freezing for 1–2 weeks. This will cause the crystals to separate from the wine and stick to the sides of the holding vessel. When the wine is drained from the vessels, the tartrates are left behind. They may also form in wine bottles that have been stored under very cold conditions. During "heat stabilization", unstable proteins are removed by adsorption onto bentonite, preventing them from precipitating in the bottled wine.

Secondary Fermentation and Bulk Aging

During the secondary fermentation and aging process, which takes three to six months, the fermentation continues very slowly. The wine is kept under an airlock to protect the wine from oxidation. Proteins from the grape are broken down and the remaining yeast cells and other fine particles from the grapes are allowed to settle. Potassium bitartrate will also precipitate, a process which can be enhanced by cold stabilization to prevent the appearance of (harmless) tartrate crystals after bottling. The result of these processes is that the originally cloudy wine becomes clear. The wine can be racked during this process to remove the lees.

The secondary fermentation usually takes place in either large stainless steel vessels with a volume of several cubic meters, or oak barrels, depending on the goals of the winemakers. Unoaked wine is fermented in a barrel made of stainless steel or other material having no influence in the final taste of the wine. Depending on the desired taste, it could be fermented mainly in stainless steel to be briefly put in oak, or have the complete fermentation done in stainless steel. Oak could be added as chips used with a non-wooden barrel instead of a fully wooden barrel. This process is mainly used in cheaper wine.

Amateur winemakers often use glass carboys in the production of their wine; these vessels (sometimes called *demijohns*) have a capacity of 4.5 to 54 litres (1.2–14.3 US gallons). The kind of vessel used depends on the amount of wine that is being made, the grapes being used, and the intentions of the winemaker.

Malolactic Fermentation

Malolactic fermentation occurs when lactic acid bacteria metabolize malic acid and produce lactic acid and carbon dioxide. This is carried out either as an intentional procedure in which specially cultivated strains of such bacteria are introduced into the maturing wine, or it can happen by chance if uncultivated lactic acid bacteria are present.

Malolactic fermentation can improve the taste of wine that has high levels of malic acid, because malic acid in higher concentration generally causes an often unpleasant harsh and bitter taste sensation, whereas lactic acid is perceived as more gentle and less sour.

The process is used in most red wines and is discretionary for white wines.

Laboratory Tests

Whether the wine is aging in tanks or barrels, tests are run periodically in a laboratory to check the status of the wine. Common tests include °Brix, pH, titratable acidity, residual sugar, free or available sulfur, total sulfur, volatile acidity and percent alcohol. Additional tests include those for the crystallization of cream of tartar (potassium hydrogen tratrate) and the precipation of heat unstable protein; this last test is limited to white wines. These tests are often performed throughout the making of the wine as well as prior to bottling. In response to the results of these tests, a winemaker can then decide on appropriate remedial action, for example the addition of more sulfur dioxide. Sensory tests will also be performed and again in response to these a wine maker may take remedial action such as the addition of a protein to soften the taste of the wine.

Brix is one measure of the soluble solids in the grape juice and represents not only the sugars but also includes many other soluble substances such as salts, acids and tannins, sometimes called Total Soluble Solids (TSS). However, sugar is by far the compound in greatest quantity and so for all practical purposes these units are a measure of sugar level. The level of sugar in the grapes is important not only because it will determine the final alcohol content of the wine, but also because it is an indirect index of grape maturity. Brix (Bx for short) is measured in grams per hundred grams of solution, so 20 Bx means that 100 grams of juice contains 20gm of dissolved compounds.

There are other common measures of sugar content of grapes, Specific gravity, Oechsle (Germany) and Beaume (France). The French Baume has the benefit that one Be° gives approximately one percent alcohol. Also one Be° is equal to 1.8 Brix, that is 1.8 grams of sugar per one hundred grams. This helps with deciding how much sugar to add if the juice is low in sugar: to achieve one percent alcohol add 1.8 grams per 100 ml or 18 grams per liter. This process is called chaptalization and is illegal in some countries (but perfectly acceptable for the home winemaker.) Generally, for the making of dry table wines a Bx of between 20 and 25 is desirable (equivalent to Be° of 11 to 14.)

A Brix test can be run either in the lab or in the field for a quick reference number to see what the sugar content is. Brix is usually measured with a refractometer while the other methods use a hydrometer. Generally, hydrometers are a cheaper alternative. For more accurate use of sugar measurement it should be remembered that all measurements are affected by the temperature at which the reading is made. Suppliers of equipment generally will supply correction charts. Volatile acidity test verifies if there is any steam distillable acids in the wine. Mainly present is acetic acid but lactic, butyric, propionic and formic acids can also be found. Usually the test checks for these acids in a cash still, but there are new methods available such as HPLC, gas chromatography and enzymatic methods. The amount of volatile acidity found in sound grapes is negligible, since it is a by-product of microbial metabolism.

It's important to remember that acetic acid bacteria require oxygen to grow. Eliminating any air in wine containers as well as a sulfur dioxide addition will limit their growth. Rejecting moldy grapes will also prevent possible problems associated with acetic acid bacteria. Use of sulfur dioxide and inoculation with a low-V.A. producing strain of Saccharomyces may deter acetic acid producing yeast. A relatively new method for removal of volatile acidity from a wine is reverse osmosis. Blending may also help—a wine with high V.A. can be filtered (to remove the microbe responsible) and blended with a low V.A. wine, so that the acetic acid level is below the sensory threshold.

Blending and Fining

Different batches of wine can be mixed before bottling in order to achieve the desired taste. The winemaker can correct perceived inadequacies by mixing wines from different grapes and batches that were produced under different conditions. These adjustments can be as simple as adjusting acid or tannin levels, to as complex as blending different varieties or vintages to achieve a consistent taste.

Fining agents are used during winemaking to remove tannins, reduce astringency and remove microscopic particles that could cloud the wines. The winemakers decide on which fining agents are used and these may vary from product to product and even batch to batch (usually depending on the grapes of that particular year). Gelatin has been used in winemaking for centuries and is recognized as a traditional method for wine fining, or clarifying. It is also the most commonly used agent to reduce the tannin content. Generally no gelatin remains in the wine because it reacts with the wine components, as it clarifies, and forms a sediment which is removed by filtration prior to bottling.

Besides gelatin, other fining agents for wine are often derived from animal and fish products, such as micronized potassium casseinate (casein is milk protein), egg whites, egg albumin, bone char, bull's blood, isinglass (Sturgeon bladder), PVPP (a synthetic compound), lysozyme, and skim milk powder.

Some aromatized wines contain honey or egg-yolk extract.

Non-animal-based filtering agents are also often used, such as bentonite (a volcanic clay-based filter), diatomaceous earth, cellulose pads, paper filters and membrane filters (thin films of plastic polymer material having uniformly sized holes).

Fermentation-Turning Grape Juice into Alcohol

Grape juice is turned into alcohol by the process of "fermentation." Grapes on the vine are covered with yeast, mold and bacteria. By putting grape juice into a container at the right temperature, yeast will turn the sugar in the juice into alcohol and carbon dioxide. The grape juice will have fermented. Fermentation is carried out in stainless steel vessels.

Yeast also gives flavor to the wine. But the yeast that is on the grape skin when it is harvested may not have the desired flavor. Other things on the outside of a grape are not good for wine (for example, acetic bacteria on the grapes can cause the wine to turn to vinegar).

The winemaker can eliminate unwanted yeast's, molds and bacteria, most commonly by using the "universal disinfectant," sulfur dioxide. Unfortunately, the sulfites which remain in the wine may cause a lot of discomfort to some wine drinkers. Some winemakers prefer NOT to do this, and purposely create wines that are subject to the vagaries (and different flavours) of the yeast that pre-exist on the grapes ("wild yeast fermentation").

The winemaker has many different yeast strains to choose from (and can use different strains at different times during the process for better control fermentation). The most common wine yeast is Saccharomyces. This is a good point to stop and mention "Brett," also known as the Brettanomyces strain of yeast (which can be added or come from wild yeast fermentation). As yeast works, it causes grape juice ("must") to get hot. But if there's too much heat, the yeast won't work. Cooling coils are necessary to maintain a temperature below 45 C. A less modern, but still wide widely used way to ferment wine is to place it in small oak barrels. "Barrel fermentation" is usually done at a lower temperature in temperature controlled rooms and takes longer, perhaps around 6 weeks. The longer fermentation and use of wood contributes to the flavour (and usually expense) of the wine.

The skins and pulp which remain in a red wine vat will rise to and float on top of the juice. This causes problems (if it dries out, it's a perfect breeding ground for injurious bacteria), so the winemaker will push this "cap" back down into the juice, usually at least twice a day. In large vats, this is accomplished by pumping juice from the bottom of the vat over the top of the cap.

Eventually the yeast is no longer changing sugar to alcohol (though different strains of yeast, which can survive in higher and higher levels of alcohol, can take over and contribute their own flavour to the wine-as well as converting a bit more sugar to alcohol).

After all this is completed what you have left is the wine, "dead" yeast cells, known as "lees and various other substances.

First Racking

After fermentation completed naturally or stopped by addition of distilled spirit, first racking is carried out. This involves the wine to stand still until most yeast cells and fine suspended material settle out. The wine is then filtered without disturbing the sediment or the yeast.

Winery Aging

The winery may then keep the wine so that there can be additional clarification and, in some wines, to give it a more complex flavours. Flavour can come from wood (or more correctly from the chemicals that make up the wood and are taken up into the wine).

The wine may be barrel aged for several months to several years. No air is allowed to enter the barrels during this period. Ignoring any additional processing that might be used, you could empty the barrels into bottles and sell your wine. However, during the winery aging, the smaller containers may develop differences. So the winemaker will probably "blend" wine from different barrels, to achieve a uniform result. Also, the winemaker may blend together different grape varieties to achieve desired characteristics.

Stabilization, Filtration

Stabilization is carried out to remove traces of tartaric acid. These tartarates present in the grape juice tend to crystallize in wine and if not removed completely can slowly reappear as glass like crystals in final bottles on storage. Stabilization with respect to tartarates may involve chilling of wine that can crystallize tartarates and these crystals can be removed by filtration.

Pasteurization

If the wine has an alcohol content less than 14% it may be heat pasteurized or cold pasteurized through microporous filters just before bottling.

Preservatives

The most common preservative used in winemaking is sulfur dioxide, achieved by adding sodium or potassium metabisulphite. Another useful preservative is potassium sorbate. Sulfur dioxide has two primary actions, firstly it is an anti microbial agent and secondly an anti oxidant. In the making of white wine it can be added prior to fermentation and immediately after alcoholic fermentation is complete. If added after alcoholic ferment it will have the effect of preventing or stopping malolactic fermentation, bacterial spoilage and help protect against the damaging effects of oxygen.

Additions of up to 100 mg per liter (of sulfur dioxide) can be added, but the available or free sulfur dioxide should be measured by the aspiration method and adjusted to 30 mg per liter. Available sulfur dioxide should be maintained at this level until bottling. For rose wines smaller additions should be made and the available level should be no more than 30 mg per liter. In the making of red wine sulfur dioxide may be used at high levels (100 mg per liter) prior to ferment to assist stabilize colour otherwise it is used at the end of malolactic ferment and performs the same functions as in white wine.

However, small additions (say 20 mg per liter) should be used to avoid bleaching red pigments and the maintenance level should be about 20 mg per liter. Furthermore, small additions (say 20 mg per liter) may be made to red wine after alcoholic ferment and before malolactic ferment to overcome minor oxidation and prevent the growth of acetic acid bacteria.

Without the use of sulfur dioxide, wines can readily suffer bacterial spoilage no matter how hygienic the winemaking practice.

Potassium sorbate is effective for the control of fungal growth, including yeast, especially for sweet wines in bottle. However, one potential hazard is the metabolism of sorbate to geraniol a potent and very unpleasant by-product. To avoid this, either the wine must be sterile bottled or contain enough sulfur dioxide to inhibit the growth of bacteria. Sterile bottling includes the use of filtration.

Filtration

Filtration in winemaking is used to accomplish two objectives, clarification and microbial stabilization. In clarification, large particles that affect the visual appearance of the wine are removed. In microbial stabilization, organisms that affect the stability of the wine are removed therefore reducing the likelihood of re-fermentation or spoilage.

The process of clarification is concerned with the removal of particles; those larger than 5–10 micrometers for coarse polishing, particles larger than 1–4 micrometers for clarifying or polishing. Microbial stabilization requires a filtration of at least 0.65 micrometers. However, filtration at this level may lighten a wines colour and body. Microbial stabilization does not imply sterility. It simply means that a significant amount of yeast and bacteria have been removed.

Bottling Wine

Producers often use different shaped bottles to denote different types of wine. Coloured bottles help to reduce damage by light. (Light assists in oxidation and breakdown of the wine into chemicals, such as mercaptan, which are undesirable.)

Cellaring Wine

Most people assume that the longer that you keep a wine, the better it will get Since its best to store wine under certain conditions, like in a cool damp underground cellar, this is known as "cellaring" wine. It is a misconception that you MUST age wine. The fact is, throughout the world, most wine is drunk "young" (that is relatively soon after it is produced, perhaps 12 to 18 months), even wines that are "better" if aged. While some wines will "mature" and become better over time, others will not and should be drunk immediately, or within a few years.

Tannin is a substance that comes from the seeds, stems and skins of grapes. Additional tannin can come from the wood during barrel aging in the winery. It is a preservative and is important to the long term maturing of wine. Through time, tannin (which has a bitter flavour) will precipitate out of the wine (becoming sediment in the bottle) and the complexity of the wine's flavour from fruit, acid and all the myriad other substances that make up the wine's character will come into greater balance.

Generally, it is red wines that are the ones that CAN (but do not have to be) produced with a fair amount of tannin with an eye towards long term storing and maturation. The bad news is that you shouldn't drink it young since it will taste too harsh (and probably cost too much, besides). The good news is that after a number of years, what you get is a prized, complex and balanced wine.

Remember that red wines get their colour from the stems and skins of the grape. This gives the wine tannin and aging capacity. White wines may have no contact with the stems and skins and will have little tannin (though some can be added, again, through barrel aging). Therefore most white wines don't age well. Even the ones which do get better through time will not last nearly as long as their red cousins. A fair average for many "ageable" whites would be about 5 to 7 years (some might go 10). On the other hand, really "ageable" reds can easily be kept for 30 years and longer.

Storing Wine

For wines that should be aged, a cellar should have proper :Temperature which does not have rapid fluctuation. 55 degrees Fahrenheit is a good, but you can live with 50 to 57 degrees Fahrenheit (10 to 14 degrees Centigrade). Wide swings in temperature will harm the wine. Having too high a temperature will age the wine faster so it won't get as complex as it might have. Having too low a temperature will slow the wine's maturation. Humidity. About 60 percent is right. This helps keep the cork moist. The wine will oxidize if the air (and its oxygen) gets to it. If the cork dries out, it can shrink and let air in. This is another reason to keep the bottles on their sides. The wine itself will help keep the cork moist.

Pectic Enzymes on Clarification of Wines

The enzymes such as pectinase that hydrolyze the large pectin molecules.

Pectin

A heavy, colloidal substance found in most ripe fruit which promotes the formation of gelatinous solutions and hazes in the finished wine. Fermenting fruit pulps with high pectin content, such as apples, should be treated with pectic enzyme, especially if the pulp is boiled to extract the fruit flavour (boiling releases the pectin, while pectic enzymes destroy it). Pectin is a structural heteropolysaccharide contained in the primary cell walls of terrestrial plants. It was first isolated and described in 1825 by Henri Braconnot. It is produced commercially as a white to light brown powder, mainly extracted from citrus fruits, and is used in food as a gelling agent particularly in jams and jellies. It is also used in fillings, sweets, as a stabilizer in fruit juices and milk drinks and as a source of dietary fibre.

Biology

In plant cells, pectin consists of a complex set of polysaccharides that are present in most primary cell walls and particularly abundant in the non-woody parts of terrestrial plants. Pectin is present not only throughout primary cell walls but also in the middle lamella between plant cells where it helps to bind cells together.

The amount, structure and chemical composition of pectin differs between plants, within a plant over time and in different parts of a plant. During ripening, pectin is broken down by the enzymes pectinase and pectinesterase; in this process the fruit becomes softer as the middle lamella breaks down and cells become separated from each other. A similar process of cell separation caused by pectin breakdown occurs in the abscission zone of the petioles of deciduous plants at leaf fall. Pectin is a natural part of human diet, but does not contribute significantly to nutrition. The daily intake of pectin from fruits and vegetables can be estimated to be around 5 g (assuming consumption of approximately 500 g fruits and vegetables per day).

In human digestion, pectin goes through the small intestine more or less intact. Pectin is thus a soluble dietary fibre.

Consumption of pectin has been shown to reduce blood cholesterol levels. The mechanism appears to be an increase of viscosity in the intestinal tract, leading to a reduced absorption of cholesterol from bile or food. In the large intestine and colon, microorganisms degrade pectin and liberate short-chain fatty acids that have positive influence on health (prebiotic effect).

Chemistry

Pectins are a family of complex polysaccharides that contain 1,4-linked a-D-galactosyluronic acid residues. Three pectic polysaccharides have been isolated from plant primary cell walls and structurally characterized. These are:

- Homogalacturonans
- Substituted galacturonans
- Rhamnogalacturonans.

Homogalacturonans are linear chains of α-(1-4)-linked D-galacturonic acid. Substituted galacturonans are characterized by the presence of saccharide appendant residues (such as D-xylose or D-apiose in the respective cases of xylogalacturonan and apiogalacturonan) branching from a backbone of D-galacturonic acid residues.

Rhamnogalacturonan I pectins (RG-I) contain a backbone of the repeating disaccharide: 4)-a-D-galacturonic acid-(1,2)-a-L-rhamnose-(1. From many of the rhamnose residues, sidechains of various neutral sugars branch off. The neutral sugars are mainly D-galactose, L-arabinose and D-xylose, the types and proportions of neutral sugars varying with the origin of pectin. Another structural type of pectin is rhamnogalacturonan II (RG-II), which is a less frequent complex, highly branched polysaccharide. Rhamnogalacturonan II is classified by some authors within the group of substituted galacturonans since the rhamnogalacturonan II backbone is made exclusively of D-galacturonic acid units.

Isolated pectin has a molecular weight of typically 60–130,000 g/mol, varying with origin and extraction conditions. In nature, around 80% of carboxyl groups of galacturonic acid are esterified with methanol. This proportion is decreased more or less during pectin extraction. The ratio of esterified to non-esterified galacturonic acid determines the Behaviour of pectin in food applications. This is why pectins are classified as high- vs. low-ester pectins – or in short HM vs. LM-pectins, with more or less than half of all the galacturonic acid esterified.

The non-esterified galacturonic acid units can be either free acids (carboxyl groups) or salts with sodium, potassium or calcium. The salts of partially esterified pectins are called pectinates, if the degree of esterification is below 5% the salts are called pectates, the insoluble acid form, pectic acid. Some plants like sugar beet, potatoes and pears contain pectins with acetylated galacturonic acid in addition to methyl esters. Acetylation prevents gel-formation but increases the stabilising and emulsifying effects of pectin.

Amidated pectin is a modified form of pectin. Here, some of the galacturonic acid is converted with ammonia to carboxylic acid amide. These pectins are more tolerant of varying calcium concentrations that occur in use. To prepare a pectin-gel, the ingredients are heated, dissolving the pectin. Upon cooling below gelling temperature, a gel starts to form. If gel formation is too strong, syneresis or a granular texture are the result, whilst weak gelling leads to excessively soft gels.

In high-ester pectins at soluble solids content above 60% and a pH-value between 2.8 and 3.6, hydrogen bonds and hydrophobic interactions bind the individual pectin chains together. These bonds form as water is bound by sugar and forces pectin strands to stick together. These form a 3-dimensional molecular net that creates the macromolecular gel. The gelling-mechanism is called a low-water-activity gel or sugar-acid-pectin gel.

In low-ester pectins, ionic bridges are formed between calcium ions and the ionised carboxyl groups of the galacturonic acid. This is idealised in the so-called "egg box-model". Low-ester pectins need calcium to form a gel, but can do so at lower soluble solids and higher pH-values than high-ester pectins. Amidated pectins behave like low-ester pectins but need less calcium and are more tolerant of excess calcium. Also, gels from amidated pectin are thermo-reversible – they can be heated and after cooling solidify again, whereas conventional pectin-gels will afterwards remain liquid.

High-ester pectins set at higher temperatures than low-ester pectins. However, gelling reactions with calcium increase as the degree of esterification falls. Similarly, lower pH-values or higher soluble solids (normally sugars) increase gelling speed. Suitable pectins can therefore be selected for jams and for jellies, or for higher sugar confectionery jellies.

Sources and Production

Apples, guavas, quince, plums, gooseberries, oranges and other citrus fruits, contain large amounts of pectin, while soft fruits like cherries, grapes and strawberries contain small amounts of pectin.

Typical levels of pectin in plants are (fresh weight):

- apples, 1–1.5%
- apricot, 1%
- cherries, 0.4%
- oranges 0.5–3.5%
- carrots approx. 1.4%
- citrus peels, 30%.

The main raw-materials for pectin production are dried citrus peel or apple pomace, both by-products of juice production. Pomace from sugar-beet is also used to a small extent. From these materials, pectin is extracted by adding hot dilute acid at pH-values from 1.5 – 3.5. During several hours of extraction, the protopectin loses some of its branching and chain-length and goes into solution. After filtering, the extract is concentrated in vacuum and the pectin then precipitated by adding ethanol or isopropanol. An old technique of precipitating pectin with aluminium salts is no longer used (apart from alcohols and polyvalent cations; pectin also precipitates with proteins and detergents).

Alcohol-precipitated pectin is then separated, washed and dried. Treating the initial pectin with dilute acid leads to low-esterified pectins. When this process includes ammonium hydroxide, amidated pectins are obtained. After drying and milling pectin is usually standardised with sugar and sometimes calcium-salts or organic acids to have optimum performance in a particular application.

Worldwide, approximately 40,000 metric tons of pectin are produced every year.

Uses

The main use for pectin is as a gelling agent, thickening agent and stabilizer in food. The classical application is giving the jelly-like consistency to jams or marmalades, which would otherwise be sweet juices. For household use, pectin is an ingredient in gelling sugar (also known as "Jam Sugar") where it is diluted to the right concentration with sugar and some citric acid to adjust pH. In some countries, pectin is also available as a solution or an extract, or as a blended powder, for home jam making.

For conventional jams and marmalades that contain above 60% sugar and soluble fruit solids, high-ester pectins are used. With low-ester pectins and amidated pectins less sugar is needed, so that diet products can be made. Pectin can also be used to stabilize acidic protein drinks, such as drinking yogurt, and as a fat substitute in baked goods. Typical levels of pectin used as a food additive are between 0.5 – 1.0% - this is about the same amount of pectin as in fresh fruit.

In medicine, pectin increases viscosity and volume of stool so that it is used against constipation and diarrhea. Until 2002, it was one of the main ingredients used in Kaopectate, along with kaolinite. Pectin is also used in throat lozenges as a demulcent. In cosmetic products, pectin acts as stabilizer. Pectin is also used in wound healing preparations and speciality medical adhesives, such as colostomy devices.

Also, it is considered a natural remedy for nausea. Pectin rich foods are proven to help nausea. In ruminant nutrition, depending on the extent of lignification of the cell wall, pectin is up to 90% digestible by bacterial enzymes. Ruminant nutritionists recommend that the digestibility and energy concentration in forages can be improved by increasing pectin concentration in the forage.

In the cigar industry, pectin is considered an excellent substitute for vegetable glue and many cigar smokers and collectors will use pectin for repairing damaged tobacco wrapper leaves on their cigars.

Pectin is also used in jellybeans.

The Effect of Pectic Enzymes in Wine Making

What are Pectins? How do they Influence the Clarification of Wine?

Broadly speaking, pectins constitute a group of closely related carbohydrates allied to such polysaccharides as starch and chemically defined as complex, methylated polygalacturonic acids. They possess colloidal properties and will gel (i.e. set to jelly in the presence of high concentrations of sugar and acid). The essential role of pectins as setting agents in jams and jellies has long been recognized. For making jam and jelly fruits which are naturally high in pectin are a decided advantage. Exactly the opposite is true in winemaking where even a small amount of pectin can cause many problems.

Although winemakers rarely encounter conditions which cause musts and wines high in pectin to gel, the fact that some is present can have other equally undesirable effects. Thus, wines containing pectin are exceptionally difficult to clear because the pectin acts as a colloid and stabilizes the haze.

The minute particles which are responsible for the wine appearing cloudy are too small to settle out naturally no matter how long a period of storage is allowed. Wines only clear because the tiny suspended particles constituting the haze are mutually attracted and combined to form larger particles whose size eventually increases sufficiently for them to begin settling under the force of gravity.

The action of protective colloid such as pectin and starch is to oppose these processes of clarification by stabilizing the haze forming particles and preventing their agglomeration. Since there is no tendency for the particles to attract each other, there is no particle growth and hence no settling so the wine remains stubbornly cloudy indefinitely.

Pectin stabilizes haze so efficiently that neither fining nor filtering is likely to be effective and may only make matters worse.

How, then, can wines whose haze forming particles are stabilized by pectin be clarified? The only satisfactory way to clear pectin clouded wines is to degrade the pectin so that the protective action on the particles causing the haze is destroyed and agglomeration can occur in the normal manner. This is accomplished by the addition of ½ teaspoon of Pectic Enzyme Powder per gallon of wine. The essential function of these enzymes is to break the long pectin chain into shorter units which are unable to act as protective colloids.

Although the addition of Pectic Enzymes can be used with great success for clearing wines with pectin stabilized hazes, it is obviously preferable to avoid encountering such difficulties at such a late stage in the production of the wine. Far less trouble would occur if the pectin could be destroyed prior to and/or during fermentation so that it is completely degraded by the time the first rack occurs. The wine would then clear more quickly and more completely. So for best results, add ½ teaspoon of Pectic Enzyme Powder to all your wines prior to the addition of yeast.

When fruit is added to beer or mead, it should also be treated with Pectic Enzymes to prevent pectin stabilized hazes.

1. Yields of free run and total yields of wine per ton of grapes have on the average been greater from enzyme treated than from untreated crushed grapes.
2. Use of pectic enzymes in crushed white grapes almost invariably deepened the colour of the wine in comparison with that of wine made from grapes to which no enzyme was added—probably owing to greater extraction of colour from the skins.
3. Colour of wines made from free run juice to which was added a pectic enzyme was in mamy cases about like that of the untreated wine, but also in about an equal number of cases the colour of the treated wines on aging became slightly deeper (yellower) than that of the untreated wines.
4. Therefore, if light colour is the primary objective, the enzyme should be added to the white juice rather than to the crushed grapes.
5. Enzyme-treated wines have invariably matured more rapidly than the untreated. The wine maker must be careful not to let the treated wines overmature before bottling. In one winery

experiment of the 1950 season the enzyme-treated wine is now inferior to the untreated in flavour and bouquet. In other cases there is little difference, or the enzyme treated is superior.

6. Enzyme-treated wines developed much more compact lees than the treated and were racked with less danger of carrying lees over into the wine.
7. As in previous seasons (1936-1949), we found that the enzyme-treated wines cleared much more rapidly than the untreated.
8. Also, they filtered much more rapidly and much more wine could be filtered with each setting of the filter.
9. Much less bentonite was needed for clarification of the enzyme-treated wines—usually 1-2 pounds per 1000 gallons as against 4-l0 pounds for the untreated.
10. At two winners relatively small amounts of Pectineal have been used to pound of Pectinol-0 per ton. The wines so treated do not clear as rapidly as with larger additions of enzyme but are found to filter very much more rapidly than the untreated and to require far less bentonite in clarification.
11. Enzyme-treated juices were found commercially to filter more readily and to give concentrates of better quality than the untreated.
12. As reported by Berg and Marsh, it was observed that Pectineal greatly increased the rate of clearing of heat extracted fermented red musts. Usually the enzyme-treated heat extracted wines in our experiments were of deeper colour than the checks, indicating that the enzyme may have stabilized the colour.
13. On the other hand, where the enzyme was used in the crushed grapes without heat extraction the colour of the resulting dry wines were in some cases less intense than in the check lots. In others it was more intense or about equal to the untreated.
14. No very clear cut differences were observed in the effect of several different Rohm and Haas enzyme preparations on the colour of either red or white wines.
15. Enzyme-treated white wines darkened more rapidly than the untreated when incubated at 45° C (131° F), although the difference was usually not very great.
16. Preliminary evidence was found to indicate that galacturonic acid formed by hydrolysis of the pectin of the must by the

enzyme may be concerned in the darkening of the colour of enzyme-treated wines.

17. Pectineal solutions of the concentrations applied in wine making did not darken; indicating that where darkening of the treated wine occurs it is some constituent of the wine itself that is concerned.
18. Ports made with Pectineal were in most cases deeper in colour than the untreated.
19. Enzyme-treated musts foamed much less than the untreated during fermentationo
20. There appeared to be no consistent differences in heat and cold stability of enzyme-treated and untreated wines. Finished enzyme-treated dry white wines were stable at 131° F and at 0° F.
21. Maintenance of a reasonable level of SO_2 in enzyme-treated white wines held the colour.

Vinegar Fermentation and Production

Vinegar is an acid liquid produced from the fermentation of ethanol in a process that yields its key ingredient, acetic acid (ethanoic acid). It also may come in a diluted form. The acetic acid concentration typically ranges from 4% to 8% by volume for table vinegar and up to 18% for pickling. Natural vinegars also contain small amounts of tartaric acid, citric acid, and other acids. Vinegar has been used since ancient times and is an important element in European, Asian, and other cuisines.

The word "vinegar" derives from the Old French *vin aigre*, meaning "sour wine", which in turn is derived from the Latin "vinum aegrum" meaning "feeble wine".

History

Vinegar has been made and used for thousands of years. Traces of it have been found in Egyptian urns dating from around 3000 BC. According to Shennong's *Herb Classic*, vinegar was invented in China during the Xia Dynasty, around 2000 BC.

In the Tanakh and the Old Testament of the Bible, it is mentioned as unpleasant to drink (Ps. 69:21) and foolish to combine with *nether* (most likely soda ash, although possibly potash, natron, or niter) (Prov. 25:20), but more favorably as a condiment when Boaz allows

Ruth to "dip her piece of bread in the vinegar" (Ruth 2:14). Jesus was offered vinegar or sour wine while on the cross (Matthew 27:48; Mark 15:36). In Islamic traditions, vinegar is one of the four favoured condiments of the Prophet Muhammad, who called it a "Blessed seasoning".

In 1864, Louis Pasteur showed that vinegar results from a natural fermentation process.

Production

Vinegar is made from the fermentation of ethanol by acetic acid bacteria. The ethanol may be derived from many different sources including wine, cider, beer or fermented fruit juice, or it may be made synthetically from natural gas and petroleum derivatives.

Commercial vinegar is produced either by fast or slow fermentation processes. Slow methods generally are used with traditional vinegars and fermentation proceeds slowly over the course of weeks or months.

The longer fermentation period allows for the accumulation of a nontoxic slime composed of acetic acid bacteria and soluble cellulose, known as the mother of vinegar. Fast methods add mother of vinegar (i.e., bacterial culture) to the source liquid before adding air using a Venturi pump system or a turbine to promote oxygenation to obtain the fastest fermentation. In fast production processes, vinegar may be produced in a period ranging from 20 hours to three days.

Varieties

Malt

Malt vinegar is made by malting barley, causing the starch in the grain to turn to maltose. Then an ale is brewed from the maltose and allowed to turn into vinegar, which is then aged. It typically is light brown in colour.

Wine

Wine vinegar is made from red or white wine and is the most commonly used vinegar in Mediterranean countries and Central Europe. As with wine, there is a considerable range in quality. Better quality wine vinegars are matured in wood for up to two years and exhibit a complex, mellow flavour. Wine vinegar tends to have a lower acidity than that of white or cider vinegars. There are more expensive wine vinegars that are made from individual varieties of wine, such as Champagne, Sherry, or pinot grigio.

Apple Cider

Apple cider vinegar, otherwise known simply as cider vinegar or ACV, is made from cider or apple must and has a brownish-yellow colour. It often is sold unfiltered and unpasteurized with the mother of vinegar present, as a natural product. It is very popular, partly because of supposed beneficial health and beauty properties. Because of its acidity, apple cider vinegar may be very harsh, even burning, to the throat. If taken straight, (as opposed to used in cooking), it can be diluted (e.g., with fruit juice or water) before drinking. It is also sometimes sweetened with sugar or honey. There have been reports of acid chemical burns of the throat from apple cider vinegar tablets, but "doubt remains as to whether apple cider vinegar was in fact an ingredient in the evaluated products."

Fruit

Fruit vinegars are made from fruit wines, usually without any additional flavouring. Common flavours of fruit vinegar include apple, black currant, raspberry, quince, and tomato. Typically, the flavours of the original fruits remain in the final product.

Most fruit vinegars are produced in Europe, where there is a growing market for high-priced vinegars made solely from specific fruits (as opposed to non-fruit vinegars which are infused with fruits or fruit flavours). Several varieties, however, also are produced in Asia. Persimmon vinegar, called *gam sikcho*, is popular in South Korea. Jujube vinegar and wolfberry vinegar are produced in China.

Umezu, a salty, sour liquid that is a by-product of *umeboshi* (pickled *ume*) production, is produced in Japan, but technically is not a true vinegar.

Jamun Sirka is a vinegar produced from the *Jamun* (or rose apple) fruit in India. It is considered to be medicinally valuable for stomach, spleen and diabetic ailments.

Balsamic

Balsamic vinegar is an aromatic, aged type of vinegar traditionally crafted in the Modena and Reggio Emilia provinces of Italy from the concentrated juice, or must, of white grapes (typically of the Trebbiano variety). It is very dark brown in colour and its flavour is rich, sweet, and complex, with the finest grades being the product of years of aging in a successive number of casks made of various types of wood (including oak, mulberry, chestnut, cherry, juniper, ash, and acacia). Originally a product available only to the Italian upper classes, a cheaper form of balsamic vinegar became widely known and available around the world in the late twentieth century.

True balsamic vinegar (which has Protected Designation of Origin) is aged for 12 to 25 years. Balsamic vinegars that have been aged for up to 100 years are available, though they are usually very expensive. The commercial balsamic sold in supermarkets is typically made with concentrated grape juice mixed with a strong vinegar, which is laced with caramel and sugar. Regardless of how it is produced, balsamic vinegar must be made from a grape product.

Balsamic vinegar has a high acidity level but the tart flavour is usually hidden by the sweetness of the other ingredients, making it very mellow.

Rice

Rice vinegar is most popular in the cuisines of East and Southeast Asia. It is available in “white” (light yellow), red, and black varieties. The Japanese prefer a light rice vinegar for the preparation of sushi rice and salad dressings. Red rice vinegar traditionally is coloured with red yeast rice. Black rice vinegar (made with black glutinous rice) is most popular in China, and it is also widely used in other east Asian countries. White rice vinegar has a mild acidity and a somewhat “flat”, uncomplex flavour. Some varieties of rice vinegar are sweetened or otherwise seasoned with spices or other added flavorings.

Coconut

Coconut vinegar, made from fermented coconut water, is used extensively in Southeast Asian cuisine (particularly in the Philippines, a major producer, where it is called *suka ng niyog*), as well as in some cuisines of India. A cloudy white liquid, it has a particularly sharp, acidic taste with a slightly yeasty note.

Palm

Palm vinegar, made from the fermented sap from flower clusters of the nipa palm (also called attap palm), is used most often in the Philippines, where it is produced, and where it is called *sukang paombong*.

Cane

Cane vinegar, made from sugar cane juice, is most popular in the Philippines, in particular, the Ilocos Region of the northern Philippines (where it is called *sukang iloko*), although it also is produced in France and the United States. It ranges from dark yellow to golden brown in colour and has a mellow flavour, similar in some respects, to rice vinegar, though with a somewhat "fresher" taste. Contrary to expectation, containing no residual sugar, it is not sweeter than other vinegars. In the Philippines, it often is labelled as *suka na maasim*, although this is simply a generic term meaning "sour vinegar."

Cane vinegars from Ilocos also varies in two different types. The "basi" or the sweet vinegar and the "Suka" which is the sour one. The sweet vinegar is used as a wine in Ilokanos, while the other type of vinegar is used as a seasoning and preservative. A white variation has become quite popular in Brazil in recent years, where it is the cheapest type of vinegar sold. It is now common for other types of vinegar (made from wine, rice and apple cider) to be sold mixed with cane vinegar to lower the costs.

Raisin

Vinegar made from raisins, called *khal 'anab* is used in cuisines of the Middle East, and is produced therein. It is cloudy and medium brown in colour, with a mild flavour.

Date

Vinegar made from dates is a traditional product of the Middle East.

Beer

Vinegar made from beer is produced in the United Kingdom, Germany, Austria, and the Netherlands. Although its flavour depends on the particular type of beer from which it is made, it is often described as having a malty taste. That produced in Bavaria, is a light golden colour with a very sharp and not-overly-complex flavour.

Honey

Vinegar made from honey is rare, although commercially available honey vinegars are produced in Italy, France, Romania and Spain.

East Asian Black

Chinese black vinegar is an aged product made from rice, wheat, millet, sorghum, or a combination thereof. It has an inky black colour and a complex, malty flavour. There is no fixed recipe and thus some Chinese black vinegars may contain added sugar, spices, or caramel colour. The most popular variety, Zhenjiang vinegar, originated in the city of Zhenjiang, in the eastern coastal province of Jiangsu, China and also is produced in Tianjin and Hong Kong.

A somewhat lighter form of black vinegar, made from rice, also is produced in Japan, where it is called *kurozu*. Since 2004 it has been marketed as a healthful drink; its manufacturers claim that it contains high concentrations of amino acids.

Flavored Vinegars

Popular fruit-flavoured vinegars include those infused with whole raspberries, blueberries, or figs (or else from flavourings derived from these fruits). Some of the more exotic fruit-flavoured vinegars include blood orange and pear.

Herb vinegars are flavoured with herbs, most commonly Mediterranean herbs such as thyme or oregano. Such vinegars can be prepared at home by adding sprigs of fresh or dried herbs to vinegar purchased at a grocery store; generally a light-coloured, mild tasting vinegar, such as that made from white wine, is used for this purpose. Sweetened vinegar is of Cantonese origin and is made from rice wine, sugar and herbs including ginger, cloves, and other spices.

Job's Tears

In Japan, an aged vinegar also is made from Job's Tears, a tall grain-bearing tropical plant. The vinegar is similar in flavour to rice vinegar.

Kombucha

Kombucha vinegar is made from kombucha, a symbiotic culture of yeast and bacteria. The bacteria produce a complex array of nutrients and populate the vinegar with bacteria which some claim promotes a healthy digestive tract, although no scientific studies have confirmed this. Kombucha vinegar primarily is used to make a vinaigrette and is flavoured by adding strawberries, blackberries, mint, or blueberries at the beginning of fermentation.

Kiwifruit

A by-product of commercial kiwifruit growing is a large amount of waste in the form of firstly misshapen or otherwise rejected fruit that may constitute up to 30 per cent of the crop and secondly kiwifruit pomace which is the presscake residue left after kiwifruit juice manufacture.

One of the uses for this waste is the production of kiwifruit vinegar. Produced commercially in New Zealand since, at least, the early 1990s Production of kiwifruit vinegar began for domestic sale in China in 2008.

Sinamak

A variation of cane vinegar from the Philippines (*sukang maasim*) is called *sinamak* which is simply a spiced version that mixes the cane vinegar with siling labuyo, onions and garlic.

Distilled Vinegar

Any type of vinegar may be distilled to produce a colourless solution of about 5% to 8% acetic acid in water. This is variously known as distilled spirit or "virgin vinegar", or white vinegar, and is used for medicinal, laboratory and cleaning purposes as well as in cooking, baking, meat preserving, including pickling. The most common starting material, because of its low cost, is malt vinegar. By FDA regulations since the 1950s, Distilled Vinegar may be synthetic ethanol made by direct chemical oxidation of wood or fossil fuels. Distilled and other highly processed vinegars are mineral deficient and, when consumed, pull calcium and other minerals from the bones and tissues.

Spirit Vinegar

The term 'spirit vinegar' is sometimes reserved for the stronger variety (5% to 20% acetic acid) made from sugar cane or from chemically produced acetic acid.

Culinary Uses

Vinegar is commonly used in food preparation, particularly in pickling processes, vinaigrettes, and other salad dressings. It is an ingredient in sauces such as mustard, ketchup, and mayonnaise. Vinegar is sometimes used while making chutneys. It is often used as a condiment. Marinades often contain vinegar.

- Condiment for beetroot — cold, cooked beetroot is commonly eaten with vinegar
- Condiment for fish and chips — People commonly use malt vinegar (or non-brewed condiment) on chips.
- Flavouring for potato chips — many American, Canadian and British manufacturers of packaged potato chips and crisps feature a variety flavoured with vinegar and salt.
- Vinegar pie — a North American dessert made with a vinegar to one's taste and similar to chess pie.
- Pickling — any vinegar can be used to pickle foods.
- Cider vinegar and sauces — cider vinegar usually is not suitable for use in delicate sauces.

- Substitute for fresh lemon juice — cider vinegar can usually be substituted for fresh lemon juice in recipes and obtain a pleasing effect although it lacks the vitamin C.
- Saucing roast lamb — pouring cider vinegar over the meat when roasting lamb, especially when combined with honey or when sliced onions have been added to the roasting pan, produces a sauce.
- Sweetened vinegar is used in the dish of pork knuckles and ginger stew which is made among Chinese people of Cantonese backgrounds to celebrate the arrival of a new child.
- Sushi rice — Japanese use rice vinegar as an essential ingredient for sushi rice.
- Red vinegar — Sometimes used in Chinese soups
- Flavouring — used in the Southern U.S. to flavour collard greens, green beans, black-eyed peas, or cabbage to taste.
- Commonly put into mint sauce, for general palate preference.
- Vinegar—especially the coconut, cane, or palm variety—is one of the principal ingredients of Philippine.

Medical Uses

Many remedies and treatments have been ascribed to vinegar over millennia and in many different cultures, however, few have been verifiable using controlled medical trials and many that are effective to some degree have significant side effects and carry the possibility of serious health risks.

Soothing for Sunburns

White vinegar applied as a spray to tissue draped over a sunburn helps restore the lost acidic level to the skin, and gives a cooling effect.

Possible Cholesterol and Triacylglycerol Effects

A 2006 study concluded that a test group of rats fed with acetic acid (the main component of vinegar) had "significantly lower values for serum total cholesterol and triacylglycerol", among other health benefits.

Rats fed vinegar or acetic acid have lower blood pressure than controls, although the effect has not been tested in humans.

Reduced risk of fatal ischemic heart disease was observed among participants in a trial who ate vinegar and oil salad dressings frequently.

Blood Glucose Control and Diabetic Management

Prior to hypoglycemic agents, diabetics used vinegar teas to control their symptoms. Small amounts of vinegar (approximately 20 ml or two tablespoons of domestic vinegar) added to food, or taken along with a meal, have been shown by a number of medical trials to reduce the glycemic index of carbohydrate food for people with and without diabetes. This also has been expressed as lower glycemic index ratings in the region of 30%.

Diet Control

Multiple trials indicate that taking vinegar with food increases satiety (the feeling of fullness) and so, reduces the amount of food consumed.

Infections

Vinegar has been used to fight infections since Hippocrates, who lived between 460-377 BC, prescribed it for curing persistent coughs. As a result, vinegar is popularly believed to be effective against infections. While vinegar can be an effective antibacterial cleaning agent on hard surfaces such as washroom tiles and countertops, studies show that vinegar – whether taken internally or applied topically – is not effective against infections, lice, or warts.

Researchers at the Food Biotechnology Department, Instituto de la Grasa (CSIC) in Seville, Spain, conducted research on the antimicrobial activity of several products. Vinegar and red and white wines were among the products tested. (Note: The focus of the research was olive oil, but it confirmed other findings related to vinegar and red and white wines.) The following microorganisms were used in the study: *S. aureus, L. monocytogenes, S. Enteritidis, E.coli* 0157:H7, *S.sonnei* and Yersinia sp. Among the items tested, vinegar (5% acetic acid) showed the strongest bactericidal activity against all strains tested, which was attributed to its high acetic acid content. The researchers noted their study confirmed previous results. It was noted that both red and white wines exhibited bactericidal activity, in particular against Salmonella Enteritidis and Yersinia sp. S. aureus and L. monocytogenes were the least sensitive to the wines.

Other Medicinal Uses

Applying vinegar to common jellyfish stings deactivates the nematocysts; however, placing the affected areas in hot water is a more effective treatment because the venom is deactivated by heat. The latter requires immersion in 45°C (113°F) water for at least four minutes for the pain to be reduced to less than what would be accomplished using vinegar. But vinegar should not be applied to Portuguese man o' war stings, however, since they are not actually jellyfish and vinegar can cause their nematocysts to discharge venom, making the pain worse. Vinegar is often used as a natural deodorant, mainly because of its antibacterial effect.

Contrary to myth, vinegar cannot be used as a detoxification agent to circumvent urinalysis testing for cannabis.

Potential Hazards

Esophageal injury by apple cider vinegar tablets has been reported, and because vinegar products sold for medicinal purposes are neither regulated nor standardized, they vary widely in content, pH, and other respects.

Long-term heavy vinegar ingestion may also cause Hypokalemia, Hyperreninemia, and Osteoporosis.

Cervical Screening Tool

Diluted vinegar 3% to 5%, has also been tested as an effective screening tool for cervical cancer. Vinegar changes the colour of affected tissue to white, making diagnosis by inspection possible, reducing in 35% the mortality for early detection against control group.

Cleaning Uses

White vinegar is often used as a household cleaning agent. Because it is acidic, it can dissolve mineral deposits from glass, coffee makers, and other smooth surfaces. For most uses dilution with water is recommended for safety and to avoid damaging the surfaces being cleaned. Vinegar is an excellent solvent for cleaning epoxy resin and hardener, even after the epoxy has begun to harden. Malt vinegar sprinkled onto crumpled newspaper is a traditional, and still-popular, method of cleaning grease-smeared windows and mirrors in the UK. Vinegar can be used for polishing brass or bronze.

Recently, vinegar has been marketed as a green solution for many household cleaning problems. For example, vinegar has been cited recently as an eco-friendly urine cleaner for pets and as a weed killer.

Agricultural and Horticultural Uses

Herbicide Use

Vinegar can be used as an herbicide. Acetic acid is not absorbed into root systems, the vinegar will kill top growth, but perennial plants will reshoot. Commercial vinegar, available to consumers for household use, does not exceed 5%, and solutions above 10% need careful handling since they are corrosive and damaging to skin. Stronger solutions that are labelled for use as herbicides are available from some retailers.

Miscellaneous

When a bottle of vinegar is opened, mother of vinegar may develop. It is considered harmless and can be removed by filtering.

Vinegar eels (*Turbatrix aceti*), a form of nematode that has cells that are air-borne, may occur in some forms of vinegar unless the vinegar is kept covered.

These feed on the mother of vinegar and can occur in naturally fermenting vinegar. This is the reason vinegar condiment jars have tightly-fitting stoppers. Most manufacturers filter and pasteurize their product before bottling to eliminate any potential adulteration, although they are harmless when ingested.

When vinegar is added to sodium bicarbonate (baking soda), it produces a volatile mixture which rapidly decomposes into water, carbon dioxide and sodium ethanoate, which makes the reaction fizz. It is often used to illustrate typical acid-base reactions in school science experiments.

Some countries prohibit the selling of vinegar over a certain percentage acidity. As an example, the government of Canada limits the acetic acid of vinegars to between 4.1% and 12.3%.

Posca, a Roman legionaries' basic drink was vinegar mixed with water and optional honey.

According to legend, in France during the Black Plague, four thieves were able to rob houses of plague victims without being infected themselves. When finally caught, the Judge offered to grant the men their freedom, on the condition that they revealed how they managed to stay healthy. They claimed that a medicine woman sold them a potion, made of garlic soaked in soured red wine (vinegar). Variants of the recipe, called Four Thieves Vinegar, have been passed down for hundreds of years and are a staple of New Orleans Voodoo practices.

Diluted vinegar can be used as a homemade stop bath during photographic processing.

7

Dairy Microbiology: Bacteria, Yeasts and Molds

Food spoilage is an enormous economic problem worldwide. Through microbial activity alone, approximately one-fourth of the world's food supply is lost. Milk is a highly nutritious food that serves as an excellent growth medium for a wide range of microorganisms. The microbiological quality of milk and dairy products is influenced by the initial flora of raw milk, the processing conditions, and post-heat treatment contamination. Undesirable microbes that can cause spoilage of dairy products include Gram-negative psychrotrophs, coliforms, lactic acid bacteria, yeasts, and molds.

In addition, various bacteria of public health concern such as Salmonella spp., Listeria monocytogenes, Campylobacter jejuni, Yersinia enterocolitica, pathogenic strains of Escherichia coli and enterotoxigenic strains of Staphylococcus aureus may also be found in milk and dairy products. For this reason, increased emphasis should be placed on the microbiological examination of milk and dairy foods.

Microbiological analyses are critical for the assessment of quality and safety, conformation with standards and specifications, and regulatory compliance.

In Hungary, a very comprehensive set of microbiological reference criteria covering virtually all foods and including limits for an extensive range of indicators and pathogens was issued by the Ministry of Health in 1998. This official document entitled "Decree No. 4/1998 (XI. 11.) EüM on the acceptable levels of microbiological contamination in foods" is based on ICMSF and FAO principles in that it includes sampling plans and identifies the criteria for acceptance or rejection of a lot. Altogether 26 food groups have mandatory criteria and the number or type of microorganisms for which there are limits for any particular food varies from one to nine. Divided into nine subgroups, milk and milk products constitute one of the 26 groups formed. As of 2003, in terms of dairy foods, this legal document was replaced by another one titled "Decree No. 1/2003 (I. 8.) FVM-ESzCsM on the food hygienic conditions for production and placing on the market of raw milk, heat-treated milk and milk-based products", which is identical in content with the EU Milk Hygiene Directive 92/46/EEC.

Production and use of Microbial Enzymes for Dairy Processing

Indians are known to be lovers of milk and its products. As a sequel to white revolution, India has surged ahead to become the largest milk producer in the world, the production figure touched to 74 million tons for the year 1998. With growing urbanization, demand for processed dairy foods has increased considerably, in particular demand for different cheese varieties, and low-lactose milk due to increasing intolerance of human beings to lactose in milk and other milk products. For improving the quality of milk and milk products, a number of different enzymes from microbial as well as from nonmicrobial sources have potential applications in dairy processing.

The use of rennet in cheese manufacture was among the earliest applications of exogenous enzymes in food processing, dating back to approximately 6000 B C. In 1994, the total production of cheese was 8000 metric tons against a total demand of 9000 metric tons. The projected demand by 2000 A D is around 30,000 metric tons. The use of rennet, as an exogenous enzyme, in cheese manufacture is perhaps the largest single application of enzymes in food processing. In recent years, proteinases have found additional applications in dairy technology, for example in acceleration of cheese ripening, modification of functional properties and preparation of dietic products. Principal among some enzymes that have important and growing applications are lipases and b-galactosidases.

Enzymes with limited applications include glucose oxidase, superoxide dismutase, sulphydryl oxidase, etc. The increasing use of enzymes to produce specific products with characteristic attributes can be emphasized by the world-wide sale of industrial enzymes approximating to US $ 1.6 billion, which is expected to reach US $ 3.0 billion by the year 2008. Almost 45% of this is shared by the food industry and the remaining is shared by detergent (34.4%), textile (11%), leather (2.8%), pulp and paper (1.2%) industries and other industries (5.6%) excluding enzymes for use in diagnostics and therapeutics.

India being the highest producer of milk in the world, and consequently the surplus availability of milk in our country has triggered the food and dairy industry to convert the liquid milk into value-added products using biochemical and enzymatic processes.

Microbial Rennets in Dairy Applications

Animal rennet (bovine chymosin) is conventionally used as a milk-clotting agent in dairy industry for the manufacture of quality cheeses with good flavour and texture. Owing to an increase in demand for cheese production world wide – i.e. 4% per annum over the past 20 years, approximating 13.533 million tons – coupled with reduced supply of calf rennet, has therefore led to a search for rennet substitutes, such as microbial rennets. At present, microbial rennet is used for one-third of all the cheese produced world wide.

Rennin acts on the milk protein in two stages, by enzymatic and by nonenzymatic action, resulting in coagulation of milk. In the enzymatic phase, the resultant milk becomes a gel due to the influence of calcium ions and the temperature used in the process. Many microorganisms are known to produce rennet-like proteinases which can substitute the calf rennet.

Microorganisms like Rhizomucor pusillus, R. miehei, Endothia parasitica, Aspergillus oryzae, and Irpex lactis are used extensively for rennet production in cheese manufacture. Extensive research that has been carried out so far on rennet substitutes has been reviewed by several authors. Different strains of species of Mucor are often used for the production of microbial rennets. Whereas best yields of the milk-clotting protease from Rhizomucor pusillus are obtained from semisolid cultures containing 50% wheat bran, R. miehei and Endothia parasitica are well suited for submerged cultivation. Using the former, good yields of milk-clotting protease may be obtained in a medium containing 4% potato starch, 3% soybean meal, and 10% barley. During growth, lipase is secreted together with the protease. Therefore, the lipase activity has to be destroyed by reducing the pH, before the preparation can be used as cheese rennet.

Microbial rennets from various microorganisms (marketed under the trade names such as Rennilase, Fromase, Marzyme, Hanilase, etc.) being marketed since the 1970s have proved satisfactory for the production of different kinds of cheese. The molecular and enzymatic properties of chymosins have been studied extensively. Although the proteolytic specificities of the three commonly used fungal rennets are considerably different from those of calf chymosin, these rennets have been used to produce acceptable cheeses.

Recently Novo Nordisk has succeeded in expressing just one proteolytic enzyme from the fungus R. miehei in the well-known organism A. oryzae. This host organism is able to produce the single protease that cleaves the -casein into a glycomacropeptide and para casein by hydrolysing only at the phe105–met106 peptide bond between phenyl alanine and methionine. This monocomponent enzyme product has the trade name Novoren. One major drawback of microbial rennet use in cheese manufacture, is the development of off flavour and bitter taste in the nonripened as well as in the ripened cheeses. The rennets from microbial sources are more proteolytic in nature in comparison to rennet from animal sources, resulting in production of some bitter peptides during the process of cheese ripening. Hence, attempts have been made to clone the gene for calf chymosin, and to express it in selected bacteria, yeasts, and molds.

Recombinant Rennets for Cheese Manufacture

Due to shortage of calf stomachs and the economic value of cheese rennet, gene for calf chymosin was one of the first genes for mammalian enzymes that was cloned and expressed in microorganisms. Many different laboratories have cloned the gene for calf prochymosin in Escherichia coli, and analysed the structure of the gene as well as the properties of the recombinant chymosin. The expressed proenzyme in E. coli, is present mainly as insoluble inclusion bodies, comprised of reduced prochymosin as well as molecules that are interlinked by disulphide bridges. After disintegration of the cells, inclusion bodies are harvested by centrifugation. The individual laboratories have reported some differences in the procedure for renaturation of prochymosin from the inclusion bodies, but all have followed the same general scheme. The enzymatic properties of recombinant E. coli chymosin are indistinguishable from those of native calf chymosin. The enzymes were identical when observed by immunodiffusion in gels, but a slight difference was observed by enzyme linked immunosorbent assay (ELISA).

The gene for prochymosin has also been cloned in Saccharomyces cerevisiae; the levels of expression have been reported to be 0.5 to 2.0% of total yeast protein. In yeast, about 20% of the prochymosin can be released in soluble form which can be activated directly; the remaining 80% is still associated with the cell debris. The cloning of chymosin was carried out without the prosequence. Though the level of chymosin mRNA was similar to that of prochymosin mRNA, no milk-clotting activity was observed in clones containing the chymosin gene only. The results suggest that the prosequence is essential for correct folding of the polypeptide chain. The zymogen for the aspartic proteases from R. pusillus, also called mucor rennin, has likewise been cloned and expressed in yeast. Studies on its conversion to active form showed that secretion of R. pusillus protease from recombinant yeast was dependent on glycosylation of the enzyme.

Compared to yeast, the filamentous fungi generally secrete larger quantities of proteins into the culture medium. Furthermore, filamentous fungi secrete the heterologous proteins with correct folding of the polypeptide chain, and process correct pairing of sulphydryl groups. The gene for R. miehei protease has been expressed in A. oryzae. Prochymosin has been expressed in Kluyveromyces lactis, A. nidulans, A. niger, and Trichoderma reesei. In most of the cases, the reported yields of the model systems were about 10–40 mg of enzyme per litre of culture medium. However, 3.3 g of enzyme per liter has also been achieved. Yeast, Kluyveromyces lactis, has recently been used as an efficient host for the secretion of recombinant chymosin, which has led to the development of large-scale production process for chymosin. If produced on an industrial scale, the yields will perhaps be of the latter magnitude. Most of the companies produce recombinant rennet of cattle calf origin in different microbial hosts, however in India, the major source of milk is buffalo, showing a different composition from that of cow. The rennet from buffalo source has inherent compatibility for clotting buffalo milk. Therefore, National Dairy Research Institute, Karnal, has taken a lead in cloning gene of buffalo chymosin. The buffalo chymosin cDNA has been cloned in E. coli and partial N-terminal sequencing of the purified buffalo chymosin indicates that it is highly homologous to the cattle chymosin.

Several cheese making experiments have been carried out with recombinant chymosin and the general aspects of recombinant chymosin have been dealt with in a report. Since most of the rennet (> 90 %) added to cheese milk is lost in the whey, immobilization would considerably extend its catalytic life. Several rennets have been immobilized, but their efficiency as milk coagulants has been questioned. So, there is a fairly general support for the view that immobilized enzymes cannot coagulate milk properly owing to inaccessability of the Phe–Met peptide bond of k-casein, and that the apparent coagulating activity of immobilized rennets is due to leaching of the enzyme from the support. Different types of conventional cheeses have been successfully made by using recombinant rennet on an experimental or pilot scale. No major differences have been detected between cheeses made with recombinant chymosins or natural enzymes, regarding cheese yield, texture, smell, taste and ripening. The recombinant chymosins are identical with calf rennet according to the report on biochemical and genetic evidences.

Lactase in Dairy Industry

Lactose, the sugar found in milk and whey, and its corresponding hydrolase, lactase or b-galactosidase, have been extensively researched during the past decade. This is because of the enzyme immobilization technique which has given new and interesting possibilities for the utilization of this sugar. Because of intestinal enzyme insufficiency, some individuals, and even a population, show lactose intolerance and difficulty in consuming milk and dairy products. Hence, low-lactose or lactose-free food aid programme is essential for lactose-intolerant people to prevent severe tissue dehydration, diarrhoea, and, at times, even death. Another advantage of lactase-treated milk is the increased sweetness of the resultant milk, thereby avoiding the requirement for addition of sugars in the manufacture of flavoured milk drinks. Manufacturers of ice cream, yoghurt and frozen desserts use lactase to improve scoop and creaminess, sweetness, and digestibility, and to reduce sandiness due to crystallization of lactose in concentrated preparations. Cheese manufactured from hydrolysed milk ripens more quickly than the cheese manufactured from normal milk.

Technologically, lactose crystallizes easily which sets limits to certain processes in the dairy industry, and the use of lactase to overcome this problem has not reached its fullest potential because of the associated high costs. Moreover, the main problem associated with discharging large quantities of cheese whey is that it pollutes the environment. But, the discharged whey could be exploited as an alternate cheap source of lactose for the production of lactic acid by fermentation. The whey permeate, which is a by-product in the manufacture of whey protein concentrates, by ultrafiltration could be fermented efficiently by Lactobacillus bulgaricus.

Lactose can be obtained from various sources like plants, animal organs, bacteria, yeasts (intracellular enzyme), or molds. Some of these sources are used for commercial enzyme preparations. Lactase preparations from A. niger, A. oryzae, and Kluyveromyces lactis are considered safe because these sources already have a history of safe use and have been subjected to numerous safety tests. The most investigated E. coli lactase is not used in food processing because of its cost and toxicity problems.

Properties of Lactase

The properties of the enzyme depend on its source. Temperature and pH optima differ from source to source and with the type of particular commercial preparation. Immobilization of the enzymes, method of immobilization, and type of carrier can also influence these optima values. In general, fungal lactase have pH optima in the acidic range 2.5–4.5, and yeast and bacterial lactases in the neutral region 6–7 and 6.5–7.5, respectively. The variation in pH optima of lactases makes them suitable for specific applications, for example fungal lactases are used for acid whey hydrolysis, while yeast and bacterial lactases are suitable for milk (pH 6.6) and sweet whey (pH 6.1) hydrolysis.

Product inhibition, e.g. inhibition by galactose, is another property which also depends on the source of lactase. The enzyme from A. niger is more strongly inhibited by galactose than that from A. oryzae. This inhibition can be overcome by hydrolysing lactose at low concentrations by using immobilized enzyme systems or by recovering the enzyme using ultrafiltration after batch hydrolysis. Lactase from Bacillus species are superior with respect to thermostability, pH operation range, product inhibition, and sensitivity against high-substrate concentration. Thermostable enzymes, able to retain their activity at 60°C or above for prolonged periods, have two distinct advantages viz. they give higher conversion rate or shorter residence time for a given conversion rate, and the process is less prone to microbial contamination due to higher operating temperature. Bacillus species have a pH optima of 6.8 and a temperature optima of 65°C. Its high activity for skim milk, and less inhibition by galactose has made it suitable for use as a production organism for lactase.

The enzymatic hydrolysis of lactose can be achieved either by free enzymes, usually in batch fermentation process, or by immobilized enzymes or even by immobilized whole cells producing intracellular enzyme. Although numerous hydrolysis systems have been investigated, only few of them have been scaled up with success and even fewer have been applied at an industrial or semi-industrial level. Several acid hydrolysis systems have been developed to industry-scale level. Large-scale systems which use free enzyme process have been developed for processing of UHT-milk and processing of whey, using

K. lactis lactase (Maxilact, Lactozyme). Several commercial immobilized systems have been developed for commercial exploitation. Snamprogretti process of industrial-scale milk processing technology in Italy is one such working systems.

They make use of fibre-entrapped yeast lactase in a batch process, and the milk used is previously sterilized by UHT. For pilot plants, there are three other processes designed and developed to handle milk;

(i) by Gist-Brocades, Rohm GmbH (Germany), and

(ii) by Sumitomo, Japan.

These are continuous processes with short residence times. Processing of whey UF-permeate is accomplished by the system developed by Corning Glass, Connecticut, Lehigh, Valio and Amerace corp. The process by Corning Glass is being applied at commercial scale in the bakers yeast production using hydrolysed whey.

Microbial Enzymes in Accelerated Cheese Ripening

Cheese ripening is a complex process mediated by biochemical and biophysical changes during which a bland curd is developed into a mature cheese with characteristic flavour, texture, and aroma. The desirable attributes are produced by the partial and gradual breakdown of carbohydrates, lipids, and proteins during ripening, mediated by several agents, viz.

(i) residual coagulants,

(ii) starter bacteria and their enzymes,

(iii) nonstarter bacteria and their enzymes,

(iv) indegenous milk enzymes, especially proteinases, and

(v) secondary inocula with their enzymes.

Proteolysis occurs in all the cheese varieties and is a prerequisite for characteristic flavour development that can be regulated by proper use of the above agents. Cheese ripening is essentially an enzymatic process which can be accelerated by augmenting activity of the key enzymes. This has the advantage of initiating more specific action for flavour development compared to use of elevated temperatures that can result in accelerating undesirable nonspecific reactions, and consequently off flavour development. Enzymes may be added to develop specific flavours in cheeses, for example lipase addition for the development of Parmesan or Blue-type cheese flavours. Attempts to accelerate the multiple secondary flavour-forming-reactions, e.g.

Strecker degradation, have been scarce. The pathways leading to the formation of flavour compounds are largely unknown, and therefore the use of exogenous enzymes to accelerate ripening is mostly an empirical process.

Studies on the hydrolysis of whole casein and isolated casein components to observe the kinetics and specificity of aspartic proteases in rennin, pepsin, and four microbial rennet substitutes indicated the considerable differences in the reaction velocity and the extent of hydrolysis on the rennet curd yield. The rennet enzymes were active even in the later phases of cheese-ripening.

Proteinases and Peptidases

Proteolysis is characteristic of most cheese varieties and is indispensable for good flavour and textural development. Proteinases used in cheese processing include (i) plasmin, (ii) rennet, and (iii) proteinases (cell wall and/or intracellular) of the starter and nonstarter bacteria. Approximately 6% of the rennet added to cheese milk remains in the curd after manufacture and contributes significantly to proteolysis during ripening. Combinations of individual neutral proteinases and microbial peptidases intensified cheese flavour, and when used in combinations with microbial rennets reduced the intensity of bitterness caused by the latter. Acid proteases in isolation cause intense bitterness. Various animal or microbial lipases gave pronounced cheese flavour, low bitterness and strong rancidity, while lipases in combination with proteinases and/or peptidases give good cheese flavour with low levels of bitterness. In a more balanced approach to the acceleration of cheese ripening using mixtures of proteinases and peptidases, attenuated starter cells or cell-free extracts (CFE) are being favoured.

Proteolytic Enzymes of Lactic Acid Bacteria in Fermented Milk Products

The proteolytic system of lactic acid bacteria is essential for their growth in milk, and contributes significantly to flavour development in fermented milk products. The proteolytic system is composed of proteinases which initially cleaves the milk protein to peptides; peptidases which cleave the peptides to small peptides and amino acids; and transport system responsible for cellular uptake of small peptides and amino acids. Lactic acid bacteria have a complex proteolytic system capable of converting milk casein to the free amino acids and peptides necessary for their growth. These proteinases include extracellular proteinases, endopeptidases, aminopeptidases, tripeptidases, and proline-specific peptidases, which are all serine proteases. Apart from lactic streptococcal proteinases, several other proteinases from nonlactostreptococcal origin have been reported. There are also serine type of proteinases, e.g. proteinases from Lactobacillus acidophillus, L. plantarum, L. delbrueckii sp. bulgaricus, L. lactis, and L. helveticus. Aminopeptidases are important for the development of flavour in fermented milk products, since they are capable of releasing single amino acid residues from oligopeptides formed by extracellular proteinase activity.

Other Dairy Enzymes

Other enzymes used for dairy food application include: (i) proteases to reduce allergic properties of cow milk products for infants, and (ii) lipases for development of lipolytic flavours in speciality cheeses. The functional properties of milk proteins may be improved by limited proteolysis through the enzymatic modification of milk proteins. An acid-soluble casein, free of off flavour and suitable for incorporation into beverages and other acid foods, has been prepared by limited proteolysis. The antigenicity of casein is destroyed by proteolysis, and the hydrolysate is suitable for use in milk-protein-based foods for infants allergic to cow milk.

Lipolysis makes an important contribution to swiss cheese flavours, due mainly to the lipolytic enzymes of the starter cultures. The characteristic peppery flavour of Blue cheese is due to short-chain fatty acids and methyl ketones. Most of the lipolysis in Blue cheese is catalysed by Penicillium roqueforti lipase, with a lesser contribution from indigenous milk lipase.

The NOVO process for production of enzyme modified cheese (EMC) uses medium-aged cheese which is emulsified, homogenized, and pasteurized, after which 'palatase' (a lipase from R. miehei) is added, with or without a proteinase, and the blend is ripened at a high temperature for one to four days. The mixture is reheated, a paste results which is suitable for inclusion in soups, dips, dressings, or snack foods. EMC technology has been developed to produce a range of characteristic cheese flavours and flavour intensities, for example swiss, blue, cheddar, provolo-nemor or romano, suitable for inclusion at low levels in many products. The claims that exogenous enzymes are effective in accelerating ripening have not led to their wide-spread use, possibly due to their high cost, difficulties in distributing them uniformly in the curd, and the possible danger of over-ripening the cheese. The other minor enzymes having limited applications in dairy processing include glucose oxidase, catalase, superoxide dismutase, sulphydryl oxidase, lactoperoxidase, and lysozymes. Glucose oxidase and catalase are often used together in selected foods for preservation. Superoxide dismutase is an antioxidant for foods and generates H_2O_2, but is more effective when catalase is present. Thermally induced generation of volatile sulphydryl groups is thought to be responsible for the cooked off-flavour in ultra high temperature (UHT) processed milk. Use of sulphydryl oxidase under aseptic conditions can eliminate this defect.

The natural inhibitory mechanism in raw milk is due to the presence of low levels of lactoperoxidase (LP), which can be activated by the external addition of traces of H_2O_2 and thiocyanate. It has been reported that the potential of LP-system and its activation enhances the keeping quality of milk. Cow milk can be provided with protective factors by the addition of lysozyme, making it suitable as an infant milk. Lysozyme acts as a preservative by reducing bacterial counts in milk without affecting the

L. bifidus *activity. The scope of application of minor enzymes to milk and milk products has been recently reviewed.*

The global market for the production of microbial enzymes for use in dairy-products manufacture is considerably large, but is being dominated only by a limited number of enzyme producers. In India the microbial dairy enzymes requirement has been very limited till now. However, with the advent of technological processes for the manufacture of different varieties of milk products, such as cheeses by the State Dairy Federations, Co-operatives and Private Dairy Product Manufacturers like Amul, Vijaya, Verka, Dynamix, Nestle, Smith Kline Beechem, etc., the markets for the sale of such products in megacities and towns has been slowly growing for the past two to three years. Presently, many of these microbial enzymes, such as microbial rennets and other enzymes, are being imported. Hence, there is a scope for the production of enzymes such as microbial rennet, lactase, proteinases, and lipases indigenously. In the near future, the requirement for these enzymes is bound to increase by leaps and bounds, basically due to requirement of value-added dairy products in the country.

Common Tests of Milk and Cream

Milk and dairy products constitute an important item of our food. These products are very suitable for microbial growth. It thus becomes necessary to know the chemistry of milk, its spoilage, method of preservation, and different dairy products where microbes play a positive rather than negative role.

Milk is considered as a complete food and it contains proteins, fat, carbohydrates, minerals, vitamins and water.

It is also a good medium for the growth of microorganisms. It is therefore, important to know the types of microorganisms present in milk, their control and use for beneficial purposes.

Milk contains relatively few bacteria when it is secreted from the udder of an healthy animal. However, during milking operations it gets contaminated from the exterior of the upper and the adjacent areas, dairy untensils, milking machines, the himds of the milkers from the soil and dust.

In this way bacteria, yeasts and molds got into the milk and constitute the normal flora of milk. The number of contaminants added from various sources depends on the care taken to avoid contamination.

The presence of these nonpathogenic organisms in milk is not serious but if these organisms multiply quickly, they can cause spoilage of milk, such as souring or putrefaction and develop undesirable odours. Control of their multiplication in milk is therefore, very essential.

Milk may also contain pathogenic organisms, derived directly from the animal or from the surroundings. Microorganisms that are harmful and found in milk are Streptococcus cremoris, Pseudomonas sp., Mycobacterium spp.

Serratia marcescens, enteric bacteria etc. Normally, milk is pasteurized before use. However, pasteurization does not kill all the bacteria; the survivors (thermodurics), depending on their initial number then multiply and if the initial number is high they cause rapid spoilage. It is imporant, therefore, that the milk be refrigerated at. around O°C soon after pasteurization to prevent the growth of these undersirable microorganisms.

Pasteurization, either at 145°F for 30 minutes or 161°P for 15"'30 seconds eliminates most of the pathogenic bacteria particularly Mycobacterium tuberculosis.

Boiling of milk destroys all microorganisms except spore formers. Sometimes, on cooling or under improper refrigeration, spores germinate and cause spoilage of boiled milk.

Detailed Descriptions: Milk Component Analysis

Analyst Performing Milk Component Test with the MilkoScan™ FT 120

This electronic instrument provides us the ability to quickly analyse milk samples for components (i.e. fat, protein, lactose, total solids and solids non-fat). MilkoScan ™ FT 120 employs the Fourier transform Infra-Red Spectrophotometer (FTIR) measuring principle. In basic terms, the instrument provides milk component results through infrared light measurement. The Foss MilkoScan ™ FT 120 can provide test results in about 30 seconds and can be calibrated for raw bovine and caprine milk, homogenized milks, cream, whey and other dairy products.

Milk Component Chemistry

Babcock Method for Fat : This is relatively inexpensive test for fat that can be performed on raw milk and a limited number of dairy products. This procedure uses sulfuric acid to "burn everything in the milk sample" except for the fat. The remaining fat in the sample is then centrifuged and determined graduated markings on specialized glassware. The procedure is limited in that individual samples can only be read to the 0.05% fat level.

Gerber Method for Fat

This is a test is close relative to the Babcock method. It is often referred to as Babcock's European cousin as this test is most common in the European dairy industry. Milk fat is separated from other milk components by adding sulfuric acid. The separation is facilitated by using amyl alcohol and centrifugation. The fat content is read directly via a special calibrated butyrometer. The procedure is limited in that individual samples can only be read to the 0.05% fat level.

Mojonnier (Ether Extract) Method

This is the chemical method of choice for total milk fat because of its inherent analytical capacity. It can provide a very precise measurement of milk fat. Because of its precision, the ether extraction procedure is the preferred chemical reference method for milk fat. Ether and alcohol are used to extract the fat from milk. The ether extract is decanted into a dry weighing dish and the ether is evaporated. The extracted fat is then dried to a constant weight and is expressed as % fat by weight.

Kjeldahl Method for Protein

This chemical method determines the nitrogen content of milk. Acid digestion of the milk sample occurs at high temperature. Nitrogen content is then converted by known formula to provide the milk protein%.

Direct Forced Air Oven Method for Total Solids

Total solids are determined by weighing milk, drying milk and weighing the dried milk residue. The sample is dried in a forced air oven for 4 hours at 100 + 1°C. The total solids content of milk is the weight of dried milk expressed as a percentage of the original milk weight.

Electronic Somatic Cell Count (ESCC) with the Fossomatic ™ Minor

Electronic instrument used in the detection of Somatic cells in milk. The milk sample is mixed with a staining solution that dyes the DNA in somatic cells. The stained sample is then pipetted into the instrument. Light is used to fluoresce the dyed somatic cells. The fluoresced cells are then counted by using a light sensitive chip.

Direct Microscopic Somatic Cell Count

The SCC is determined by viewing stained somatic cells under a microscope under oil immersion. The analysts' estimated counts are determined by using a microscopic factor based on the field of view of the microscope.

Freezing Point Depression (FPD) of Milk with the Cryoscope

This test is used to detect the adulteration of water in milk. Due to the soluble components of milk (i.e. lactose and salts), normal milk freezes slightly below the "freezing point" of pure water. The cryoscope instrument is used to determine the freezing point of milk. Small amounts of added water dilute the soluble components of milk causing the freezing point to move upward toward that of pure water.

Interpretations of cryoscope readings can be awkward. Years ago, cryoscope readings were reported with a unique measurement scale known at the Hortvet scale (°H). However, today most cryoscope results are reported in Centigrade (°C). In some cases, laboratories have inadvertently calibrated their cryoscopes to °H and reported results as °C. When interpreting cryoscope results, first determine which measurement scale (°H or °C), is being used to calibrate the lab's cryoscope. Then apply the reading to the appropriate scale. Normal comingled loads of milk should have cryoscope values between-0.517 and-0.527 °C (-0.540 and-0.550 °H).

However, normal single herd result may vary depending on a variety of conditions including (but not limited to) herd health, season of the year, mineral content of the herd's diet, animal breed and herd stress. Because of these factors, the concerns implicating adulteration of milk with water usually are noted at-0.503 °C (-0.525 °H). If a herd's cryoscope results consistently show concern, an evaluation of the herd's milk should be conducted to determine a "baseline" for that specific herd's normal cryoscope reading. The cryoscope at Regulatory Services' milk lab is calibrated in °C and all test results are reported in °C.

Standard Plate Count Bacteria (SPC) with Petrifilm®

The Petrifilm® plate is an all-in-one plating system method that provides for the Standard Plate Count (SPC) of aerobic bacteria in milk. Pre-made films of agar are inoculated with 1:100 and 1:1000 dilutions of milk. The inoculated Petrifilm® is incubated for 48 hours and the bacteria colonies are counted by the analyst.

Antibiotic Test with Charm SL®

Detection of antibiotics or other growth inhibitors in raw milk is made possible by validated growth inhibitor assays or by rapid receptor based screening processes. Acceptable test methods have been validated and approved by FDA are identified in PMO related documents. The Charm SL® is approved for the determination of antibiotics in raw cow and goat milk. The Charm SL® is a receptor assay utilizing rapid one step assay lateral flow technology. The procedure detects Beta-Lactam drugs (Amoxicillin, Ampicillin, Ceftiofor, Cephapicin, Penicillin G) at or below safe level/tolerance.

Sediment

The sediment test determines the level of particulates present in milk and is an indicator of milk production practices and cleanliness. Sediment is determined by filtering a portion of milk by suction through a fine fibre filter. The filter is then visually examined to determine the milligrams of sediment present per gallon. In years past, the sediment test was quite common throughout the dairy industry. Improvements in farm production practices and in milk quality have diminished the need for routine sediment testing in most parts of the US.

Other Tests

The milk laboratory at Regulatory Services has the capacity to perform the milk analyses identified above. Periodically we may add new procedures. Please feel free to contact us for milk testing inquiries not identified in this document.

Milk Testing and Quality Control

Milk testing and quality control is an essential component of any milk processing industry whether small, medium or large scale. Milk being made up of 87% water is prone to adulteration by unscrupulous middlemen and unfaithful farm workers. Moreover, its high nutritive value makes it an ideal medium for the rapid multiplication of bacteria, particularly under unhygienic production and storage at ambient temperatures. We know that, in order for any processor to make good dairy products, good quality raw materials are essential.

A milk processor or handler will only be assured of the quality of raw milk if certain basic quality tests are carried out at various stages of transportation of milk from the producer to the processor and finally to the consumer. There are a number of standard manuals and text books on milk quality control. However these may not be easily available to the emerging small scale to medium scale processors in Kenya.

For these reasons, the Training Programme for the Small Scale Dairy Sector under project GOK/FAO/TCP/KEN/6611, has assembled this guide on Milk Testing and Quality Control so that it may be used for training and by the private small scale dairy processors. The methods selected are simple and basic and will suffice the requirements of most milk quality control laboratories of small scale processing units. For the larger plants with bigger laboratories more tests are to be found in the bibliography at the end of this booklet.

What is Milk Quality Control

Milk quality control is the use of approved tests to ensure the application of approved practices, standards and regulations concerning the milk and milk products. The tests are designed to ensure that milk products meet accepted standards for chemical composition and purity as well as levels of different micro-organisms.

Why have Milk Quality Control

Testing milk and milk products for quality and monitoring that MILK PRODUCTS, PROCESSORS and MARKETING AGENCIES adhere to accepted codes of practices costs money. There must be good reasons why we have to have a quality control system for the dairy industry in Kenya.

The reasons are:

i) To the Milk Producer. The milk producer expects a fair price in accordance with the quality of milk she/he produces.

ii) The Milk Processor. The milk processor who pays the producer must assure himself/herself that the milk received for processing is of normal composition and is suitable for processing into various dairy products.

iii) The Consumer. The consumer expects to pay a fair price for milk and milk products of acceptable to excellent quality.

iv) The Public and Government Agencies. These have to ensure that the health and nutritional status of the people is protected from consumption of contaminated and sub-standard foodstuffs and that prices paid are fair to the milk producers, the milk processor and the final consumer.

All the above-is only possible through institution of a workable quality testing and assurance system conforms to national or internationally acceptable standards.

Quality Control in the Milk Marketing Chain in Kenya

i) At the farm. Quality control and assurance must begin at the farm. This is achieved through farmers using approved practices of milk production and handling; and observation of laid down regulations regarding, use of veterinary drugs on lactating animals, regulations against adulterations of milk etc.

ii) At Milk collection Centres. All milk from different farmers or bulked milk from various collecting centres must be checked for wholesomeness, bacteriological, and chemical quality.

iii) At the Dairy Factories. Milk from individual farmers or bulked milk from various collecting centres

iv) Within the Dairy Factory. Once the dairy factor has accepted the farmer milk it has the responsibility of ensuring that the milk is handled hygienically during processing. It must carry out quality assurance test to ensure that the products produced conform to specified standards as to the adequacy of effect of processes applied and the keeping quality of manufactured products. A good example is the phosphatase test used on pasteurised milk and the acidity development test done on U.H.T milk.

v) During marketing of processed products. Public Health authorities are employed by law to check the quality of food stuffs sold for public consumption and may impound substandard or contaminated foodstuffs including possible prosecution of culprits. This is done in order to protect the interest of the milk consuming public.

Techniques Used in Milk Testing and Quality Control

Milk Sampling

Accurate sampling is the first pre-requisite for fair and just quality control system. Liquid milk in cans and bulk tanks should be thoroughly mixed to disperse the milk fat before a milk sample is taken for any chemical control tests. Representative samples of packed products must be taken for any investigation on quality. Plungers and dippers me used in sampling milk from milk cans.

Sampling Milk for Bacteriological Testing

Sampling milk for bacteriological tests require a lot of care. Dippers used must have been sterilised in an autoclave or pressure cooker for at least 15mm at 120° C before hand in order not to contaminate the sample. On the spot sterilisation may be employed using 70% Alcohol swab and flaming or scaling in hot steam or boiling water for 1 minute.

Preservation of Sample

Milk samples for chemical tests: Milk samples for butterfat testing may be preserved with chemicals like Potassium dichromate (1 Tablet or ½ ml 14% solution in a ¼ litre sample bottle is adequate.) Milk samples that have been kept cooling a refrigerator or ice-box must first be warmed in water bath at 40 °C, cooled to 20°C, mixed and a sample then taken for butterfat determination. Other preservative chemicals include Sodium azid at the rate of 0.08% and Bronopol (2-bromo-2-nitro-1,3-propanediol) used at the rate of 0.02%.

If the laboratory cannot start work on a sample immediately after sampling, the sample must be cooled to near freezing point quickly and be kept cool till the work can start. If samples are to be taken in the field e.g. at a milk cooling centre, ice boxes with ice pecks are useful.

Labelling and Records Keeping

Samples must be clearly labelled with name of farmer or code number and records of dates, and places included in standard data sheets. Good records must be kept neat and in a dry place. It is desirable that milk producers should see their milk being tested, and the records should be made available to them if they so require.

Common Testing of Milk

Organoleptic Tests

The organoleptic test permits rapid segregation of poor quality milk at the milk receiving platform. No equipment is required, but the milk grader must have good sense of sight, smell and taste. The result of the test is obtained instantly, and the cost of the test are low. Milk which cannot be adequately judged organoleptically must be subjected to other more sensitive and objective tests.

Procedure:

- Open a can of milk.
- Immediately smell the milk.
- Observe the appearance of the milk.
- If still unable to make a clear judgement, taste the milk, but do not swallow it. Spit the milk sample into a bucket provided for that purpose or into a drain basin, flush with water.
- Look at the can lid and the milk can to check cleanliness.

Abnormal smell and taste may be caused by:

- Atmospheric taint (e.g. barny/cowy odour).
- Physiological taints (hormonal imbalance, cows in late lactation-spontaneous rancidity).
- Bacterial taints.
- Chemical taints or discolouring.
- Advanced acidification (pH < 6.4).

Clot on Boiling (C.O.B) Test

The test is quick and simple. It is one of the old tests for too acid milk (pH<5.8) or abnormal milk (e.g. colostral or mastitis milk). If a milk sample fails in the test, the milk must contain many acid or rennet producing microrganisms or the milk has an abnormal high percentage of proteins like colostral milk. Such milk cannot stand the heat treatment in milk processing and must therefore be rejected.

Procedure: Boil a small amount of milk in a spoon, test tube or other suitable container. If there is clotting, coagulation or precipitation, the milk has failed the test. Heavy contamination in freshly drawn milk cannot be detected, when the acidity is below 0.20-0.26% Lactic acid.

The Alcohol Test

The test is quick and simple. It is based on instability of the proteins when the levels of acid and/or rennet are increased and acted upon by the alcohol. Also increased levels of albumen (colostrum milk) and salt concentrates (mastitis) results in a positive test.

Procedure: The test is done by mixing equal amounts of milk and 68% of ethanol solution in a small bottle or test tube. (68 % Ethanol solution is prepared from 68 mls 96% (absolute) alcohol and 28 mls distilled water). If the tested milk is of good quality, there will be no coagulation, clotting or precipitation, but it is necessary to look for small lumps. The first clotting due to acid development can first be seen at 0.21-0.23% Lactic acid. For routine testing 2 mls milk is mixed with 2 mls 68% alcohol.

The Alcohol-Alizarin Test

The procedure for carrying out the test is the same as for alcohol test but this test is more informative. Alizarin is a colour indicator changing colour according to the acidity. The Alcohol Alizarin solution can be bought ready made or be prepared by adding 0.4 grammes alizarin powder to 1 litre of 61% alcohol solution.

Acidity Test

Bacteria that normally develop in raw milk produce more or less of lactic acid. In the acidity test the acid is neutralised with 0.1 N Sodium hydroxide and the amount of alkaline is measured. From this, the percentage of lactic acid can be calculated. Fresh milk contains in this test also "natural acidity" which is due to the natural ability to resist pH changes. The natural acidity of milk is 0.16-0.18%.

Apparatus:

- A porcelain dish or small conical flask
- 10 ml pipette, graduated
- 1 ml pipette
- A Burette, 0.1 ml graduations
- A glass rod for stirring the milk in the dish

- A Phenophtalein indicator solution, 0.5% in 50% Alcohol
- N Sodium hydroxide solution.

Procedure: 9 ml of the milk measured into the porcelain dish/ conical flask,1 ml Phenopthalein is added and then slowly from the burret, 0.1 N Sodium hydroxide under continuous mixing, until a faint pink colour appears.

The number of mls of Sodium hydroxide solution divided by 10 expresses the percentage of lactic acid.

Resazurin Test

Resazurin test is the most widely used test for hygiene and the potential keeping quality of raw milk. Resazurin is a dye indicator. Under specified conditions Resazurin is dissolved in distilled boiled water. The Resazurin solution can later be used to test the microbial activity in a given milk sample.

Resazurin can be carried out as:

i. 10 min test.

ii. 1 hr test.

iii. 3 hr test.

The 10 min Resazurin test is useful and rapid, screening test used at the milk platform.

The 1 hr test and 3 hr tests provide more accurate information about the milk quality, but after a fairy long time. They are usually carried out in the laboratory.

Apparatus and reagents:

- Resazurin tablets
- Test tubes with 10 mls mark
- 1 ml pipette or dispenser for Resazurin solution.
- Water bath thermostatically controlled
- Lovibond comparator with Resazurin disc 4/9.

Procedure: The solution of Resazurin as prepared by adding one tablet to 50 mls of distilled sterile water. Rasazurin solution must not be exposed to sunlight, and it should not be used for more than eight hours because it losses strength.

Mix the milk and with a sanitized dipper put 10 mls milk into a sterile test tube.

Add one ml of Resazurin solution, stopper with a sterile stopper, mix gently the dye into the milk and mark the tube before the incubation in a water bath, place the test tube in a Lovibond comparator with Resazurin disk and compare it colourimetrically with a test tube containing 10 ml milk of the same sample, but without the dye (Blank).

The Gerber Butterfat Test

The fat content of milk and cream is the most important single factor in determining the price to be paid for milk supplied by farmers in many countries.

Also, in order to calculate the correct amount of feed ration for high yielding dairy cows, it is important to know the butterfat percentage as well as well as the yield of the milk produced. Further more the butterfat percentage in the milk of individual animals must be known in many breeding programmes. Butterfat tests are also done on milk and milk products in order to make accurate adjustments of the butterfat percentage in standardised milk and milk products.

Apparatus for DF test:

- Gerber butyrameters, 0-6% or 0-8% BF
- Rubber stoppers for butyrometers
- 10.94 or 11 ml pipettes for milk
- 10 mls pippetes or dispensers for Gerber Acid
- 1 mls pippetes or dispensers for Amyl alcohol
- stands for butyrometers.

Gerber water bath Reagents:

- Gerber sulphuric acid,(1.82 g/cc)
- Amyl alcohol.

Treatment of Samples

Fresh milk at approximately 20°C should be mixed well. Samples kept cool for some days should be warmed to 40°C, mixed gently and cooled to 20°C before the testing.

Procedure: Add 10 mIs sulphuric acid to the butyrometer followed by 10.94 or 11 mls of well mixed milk. Avoid wetting of the neck of the butyrometer.

Next add 1 ml of Amyl alcohol, insert stopper and shake the butyrometer carefully until the curd dissolves and no white particles can be seen. Place the butyrometer in the water bath at 65°C and keep it there until a set is ready for centrifuging. The butyrometer must be placed in the centrifuge with the stem (scale) pointing towards the centre of the centrifuge.

Spin for 5 min. at 1100 rpm.

Remove the butyrometers from the centrifuge.

Put the butyrometers in a water bath maintained at 65°C for 3 min. before taking the reading.

(Note: When transferring the butyrometers from the centrifuge into the water bath make sure that the butyrometers are all the time held with the NECK POINTING UP).

The fat column should be read from the lowest point of the meniscus of the interface of the acid-fat to the 0-mark of the scale and read the butterfat percentage.

The butyrometers should be emptied into a special container for the very corrosive liquid of acid-milk, and the butyrometers should be washed in warm water and dried before the next use.

Appearance of the Test:

- The colour of the fat column should be straw yellow.
- The ends of the fat column should be clearly and sharply defined.
- The fat column should be free from specks and sediment.
- The water just below the fat column should be perfectly clear.
- The fat should be within the graduation.

Problems in Test Results

Curdy tests:

- Too lightly coloured or curdy fat column can be due to:
- Temperature at milk or acid or both too low.
- Acid too weak.
- Insufficient acid.
- Milk and acid not mixed thoroughly.

Charred tests:

- Darkened fat column containing black speck at the base is due to:
- Temperature of milk-acid mixture too high.
- Acid too strong.
- Milk and acid mixed too slowly.

- Too much acid used.
- Acid dropped through the milk.

The Lactometer test

Addition of water to milk can be a big problem where we have unfaithful farm workers, milk transporters and greedy milk hawkers. A few farmers may also fall victim of this illegal practice. Any buyer of milk should therefore assure himself/herself that the milk he/she purchases is wholesome and has not been adulterated. Milk has a specific gravity. When its adultered with water or other materials are added or both misdeeds are committed, the density of milk change from its normal value to abnormal.

The lactometer test is designed to detect the change in density of such adulterated milk. Carried out together with the Gerber butterfat test, it enables the milk processor to calculate the milk total solids (% TS) and solids not fat (SNF). In normal milk SNF should not be below 8.5% according to Kenya Standards(KBS No 05-10:-1976).

Procedure: Mix the milk sample gently and pour it gently into a measuring cylinder (300-500). Let the Lactometer sink slowly into the milk. Read and record the last Lactometer degree (°L) just above the surface of the milk. If the temperature of the milk is different from the calibration temperature (Calibration temperature may be=20 0C) of the lactometer, calculate the temperature correction. For each °C above the calibration temperature add 0.2°L; for each °C below calibration temperature subtract 0.2 °L from the recorded lactometer reading.

For the calculations, use lactometer degrees, and for the conversion to density write 1.0 in front of the true lactometer reading,i.e. 1.030 g/ml. Clever people may try to adulterate milk in such a way that the lactometer cannot show the adulteration. But look to see if there is an unusual sediment from the milk at the bottom of the milk can and taste to find out if the milk is too sweet or salty to be normal. Samples of milk from individual cows often have lactometer reading outside the range of average milk, while samples of milk from herds should have readings hear the average milk, but wrong feeding, may result in low readings. Kenyan standards expects milk to have specific gravity of 1.026-1.032 g/ml which implies a Lactometer reading range of 26.0-32.0 °L. If the reading is consistently lower than expected and the milk supplier disputes any wrong doing arrange to take a genuine sample from the supplier (i.e. inspect milk right from source).

Freezing Point Determination

The freezing point of milk is regarded to be the most constant of all measurable properties of milk. A small adulteration of milk with water will cause a detectable elevation of the freezing point of milk from its normal values of-0.54°C. Since the test is accurate and sensitive to added water in milk, it is used to detect whether milk is of normal composition and adulterated.

Inhibitor Test

Milk collected from producers may contain drugs and/or pesticides residues. These when present in significant amounts in milk may inhibit the growth of lactic acid bacteria used in the manufacture of fermented milk such as Mala, cheese and Yoghurt, besides being a health hazard.

Principle of the Method: The suspected milk sample is subjected to a fermentation test with starter culture and the acidity checked after three (3) hours. The values of the titratable acidity obtained is compared with titratable acidity of a similarly treated sample which is free from any inhibitory substances.

Materials:

- test tubes
- Starter culture
- lml pipette
- water bath
- material for determination of titratable acidity.

Procedure: Three test tubes are filled with 10 ml of sample to be tested and three test tubes filled with normal milk.

All tubes are heated to 90 0C by putting them in boiling water for 3-5 minutes.

After cooling to optimum temperature of the starter culture (30,37, or 42°C), 1 ml of starter culture is added to each test tube, mixed and incubated for 3 hours.

After each hour, one test tube is from the test sample and the control sample is determined.

Assessment of results:

- If acid production in suspected sample is the same as the normal sample, then the suspect sample does not contain any inhibitory substances;

- If acid production as suspect sample is less than in the normal milk sample, then, the suspect sample contains antibiotics or other inhibitory substances.

Quality Control of Pasteurised Milk

When milk is pasteurised at 63°C for 30 min in batch pasteuriser or 72°C for 15 seconds in heat exchanger, continuous flow pasteurisers, All Pathogenic Bacteria Are Destroyed, there by rendering milk safe for human consumption. Simultaneously various enzymes present in milk, and which might affect its flavour, are destroyed.

In order to determine whether or not milk has been adequately pasteurised, one of the enzymes normally present in milk phosphatase, is measured. A negative phosphatase result indicates that the enzyme and any pathogenic bacteria have been destroyed during pasteursation. If it is positive, it means the pasteurisation process was inadequate and the milk may not be safe for human consumption and will have a short shelf life.

- Test tubes
- 5 mls pipettes
- 1 ml pipettes
- 100 ml volumetric flask
- 500 ml volumetric flask
- water bath at 37°C.

Note: All glassware must be rinsed, cleaned, rinsed in chromic acid solution and boiled in water for 30 min.

Reagent

Buffer Solution: Is mixed by 0.75g anhydrous sodium carbonate and 1.75g Sodium bicarbonate in 500 ml distilled water.

Buffer-substrate Solution: Place 0.15 g of di-sodium paranitrophenylphosphate (the substrate) into a clean 100ml measuring cylinder.

Add the buffer solution to make to 100 ml mark.

Store this buffer-substrate solution in a refrigerator and protected against light. It should not be used after one week. Prepare a fresh stock.

Procedure: Pipette 5mls buffer-substrate solution into a test tube, stopper and warm the solution in the water bath at 37°C. Add to the test tube 1ml of the milk to be tested, stopper and mix well and place in water bath at 37°C. Prepare a blank sample from boiled milk of the same type as that undergoing the test. Incubate both the test samples and the blank sample at 37°C for 2hrs. After incubation, remove the tubes and mix them thoroughly.

Place one sample against the blank in a Lovibond comparator" All Purposes" using A.P.T.W. disc and rotate the disc until the colour of the test sample is matched and read the disc number.

Interpretation:

Disc Reading after 2 hrs incubationat 37°C	*Remarks*
0-10	Properly pasteurised
10-18	Slightly under pasteurised
18-42	UNDER PASTEURISED
> 42	NOT PASTEURISED

Common Adulterants/Contaminants in Food and Simple Screening Tests for their Detection

Adulteration in food is normally present in its most crude form, prohibited substances are either added or partly or wholly substituted. In India normally the contamination/adulteration in food is done either for financial gain or due to carelessness and lack in proper hygienic condition of processing, storing, transportation and marketing. This ultimately results that the consumer is either cheated or often become victim of diseases. Such types of adulteration are quite common in developing countries or backward countries. However, adequate precautions taken by the consumer at the time of purchase of such produce can make him alert to avoid procurement of such food. It is equally important for the consumer to know the common adulterants and their effect on health.

8

Dairy Wastes and Treatment

The rapid growth in the size of dairy operations has resulted in new laws and regulations governing the handling and disposal of manure (Mitchell and Beddoes 2000). Requirements for nutrient management plans, manure solids disposal, and odour control (House Bill 2001) make it necessary that new manure management approaches be considered. One of the more promising methods is anaerobic digestion. Anaerobic digestion is a natural process that converts biomass to energy. Biomass is any organic material that comes from plants, animals or their wastes.

Anaerobic digestion has been used for over 100 years to stabilize municipal sewage and a wide variety of industrial wastes. Most municipal wastewater treatment plants use anaerobic digestion to convert waste solids to gas. The anaerobic process removes a vast majority of the odorous compounds (Lusk 1995),(Wilkie 2000),(Wilkie 2000). It also significantly reduces the pathogens present in the slurry (Lusk 1995). Over the past 25 years, anaerobic digestion processes have been developed and applied to a wide array of industrial and agricultural wastes (Speece 1996), (Ghosh 1997). It is the preferred waste treatment process since it produces, rather than consumes, energy and can be carried out in relatively small, enclosed tanks. The products of anaerobic digestion have value and can be sold to offset treatment costs (Roos 1991).

Anaerobic digestion provides a variety of benefits. The environmental benefits include:

- Odors are significantly reduced or eliminated.
- Flies are substantially reduced.

- A relatively clean liquid for flushing and irrigation can be produced.
- Pathogens are substantially reduced in the liquid and solid products.
- Greenhouse gas emissions are reduced.
- And finally, nonpoint source pollution is substantially reduced. On the economic side, additional benefits are provided.
- The time devoted to moving, handling, and processing manure is minimized.
- Biogas is produced for heat or electrical power.
- Waste heat can be used to meet the heating and cooling requirements of the dairy.
- Concentrating nutrients to a relatively small volume for export from the site can reduce the land required for liquid waste application.
- The rich fertilizer can be produced for sale to the public, nurseries, or other crop producers.
- Income can be obtained from the processing of imported wastes (tipping fees), the sale of organic nutrients, greenhouse gas credits, and the sale of power.
- Power tax credits may be available for each kWh of power produced.
- Greenhouse tax credits may become available for each ton of carbon recycled.
- Finally the power generated is "distributed power" which minimizes the need to modify the power grid. The impact of new power on the power grid is minimized.

In order to achieve the benefits of anaerobic digestion, the treatment facility must be integrated into the dairy operation. Unfortunately, no single dairy can serve as a model for a manure treatment facility. The operation of the dairy will establish the digester loading and the energy generated from the system. The anaerobic facility must be designed to meet the individual characteristics of each dairy.

This manual provides an introduction to the anaerobic digestion of dairy manure. It is divided into three parts. The first describes the operation and waste management practices of Idaho dairies. The second introduces anaerobic digestion and the anaerobic digestion processes suitable for dairy waste. The third presents typical design applications for different types of dairies and establishes the cost and benefits of the facilities.

Dairy Operations

Dairy operations significantly affect the quantity and quality of manure that may be delivered to the anaerobic digestion system. In addition to the number of milk and dry cows, the housing, transport, manure separation, and bedding systems used by the dairy establishes the amount of material that must be handled and the amount of energy produced.

Housing System

Confined dairy animals may be housed in a variety of systems. Commonly used housing systems include free stalls, corrals with paved feed lanes, and open lot systems. Milk cows, dry cows and heifers may be housed in free stalls, corrals, and open-lots on the same dairy. The type of housing used determines the quantity of manure that can be economically collected.

Free Stall Barns

Free stalls are currently the most popular method for housing large dairy herds. Free stall housing provides a means for collecting essentially all of the manure.

Corrals

Corral systems with paved feed lanes are also commonly used. The manure deposited in the feed lanes can be scraped or flushed daily. From 40 to 55- percent of the excreted manure may be deposited and collected from the corral feed lane. The balance of the manure may be deposited in the milk barn (10 to 15 percent) or the open lot (30 to 50%). Typically the manure deposited in the open lot is removed two to three times a year. It may have little net energy value after being stored in the open lot over prolonged periods of time.

For corral systems one must make a reasonable determination of the recoverable manure deposited in the feed lane, corral, and milk barn. Corral with Paved Feed Lane for Scrape or Vacuum Collection Corral systems also use a considerable amount of bedding material during the winter months. The straw bedding is generally removed in the spring and placed on the fields prior to spring planting.

Milk Barn

Dairy cows are milked two to three times a day. The cows are moved from their stalls to the milk parlor holding area. The milk parlor and holding area are normally flushed with fresh water. From 10 to 15 percent of the manure is deposited in the milk parlor. In addition to the manure that is flushed, the cows may be washed with a sprinkler system. Warm water that is produced by the refrigeration compressors, vacuum pumps, and milk cooling system may be used for drinking water, manure flushing or washing the cows. It has been estimated that 5 to 150 gallons of fresh water per milk cow is used in the milking centre. More common values are 10 to 30 gallons of fresh water per milk cow. The quantity and quality of water discharge from the milk parlor must be accurately measured. In many cases, the waste deposited in the milk barn is processed in a separate waste management system.

Open Lot

In open lot systems the manure is deposited on the ground and scraped into piles. The manure is removed infrequently (once or twice a year). A significant amount of manure degradation occurs resulting in greenhouse gas emissions. In many cases, the open lot degradation produces manure that has little or no net energy value.

Open Lot System

Transport System

The commonly used manure transport systems are flush, scrape, vacuum, and loader systems. In free stall barns the manure can be flushed, scraped, or vacuum collected.

Flush Systems

If a flush system is used the manure is substantially diluted. The quantity of water used in a flush system depends on the width, length, and slope of the flush isle. The feed isles are generally 14 feet wide while the back isles are generally 10 feet wide. The slope varies between one and two percent. A flush system will generally reduce the concentration of manure from 12 1/2 percent solids, “as excreted”, to less than one percent solids in the flush water. Flush systems are however more economical and less labour-intensive than scrape or vacuum systems.

Scrape Systems

Scrape systems are simply systems that collect the manure by scraping it to a sump. Under normal weather conditions the scraped manure has approximately the same consistency as the “as excreted” manure. During the warm dry summer manure may be dewatered on the slab.

Front End Loader

Front-end loaders are used to stack and remove corral bedding and manure.

Vacuum Systems

Vacuum systems collect "as excreted" manure with a vacuum truck. Generally, the trucks collect approximately 4000 gallons per load. The manure can be hauled to a disposal site rather than to an intermediate sump. Vacuum collection is a slow and tedious process. The advantage is that the collected manure is undiluted and approximately equal to the "as excreted" concentration.

Bedding

The type of bedding used can significantly alter the characteristics of the manure being treated. Typically straw, wood chips, sand, or compost are used as bedding material. In some cases paper mixed with sawdust is used. Compost usually has some sand mixed with the organic constituents. If composting is carried out on dirt lots, a significant amount of sand and silt may be incorporated into the compost. Since anaerobic digestion will not degrade the wood chips, sand, or silt, it is necessary to remove those constituents prior to, or during anaerobic digestion process. The quantity of non-degradable, organic and inorganic material can significantly impact the performance of the anaerobic digester.

The quantity of bedding added to the manure is a function of the design and operation of the dairy. Generally only the "kick-out" from the stalls is added to the manure. The quantity that is "kicked-out" is a function of the design of the dairy housing system as well as the type of bedding used.

Manure Processing

Each dairy has its own manure processing system. Scraped or flushed manure may be processed in a system separate from the milk barn waste, or the collected manure waste may be processed with the milk barn waste. In general, current manure processing consists of macerating the waste with a chopper pump, screening the waste to remove the organic fibers, followed by sedimentation to remove the sand, silt, and organic settable particles. Much of the degradable manure is removed during the separation processes. Up to 80% of the COD and 30% of the total Nitrogen and Phosphorous can be lost in the solids removed by the screen and sedimentation process. Detailed sampling and analysis is required to confirm losses.

Holding Tanks and Chopper Pumps

A wide variety of holding tanks and chopper pumps are used throughout the dairy industry. Typically, the tanks are relatively small but in some cases they are designed to hold several hours of flush water.

Anaerobic Digestion

Anaerobic digestion is the breakdown of organic material by a microbial population that lives in an oxygen free environment. Anaerobic means literally "without air". When organic matter is decomposed in an anaerobic environment the bacteria produce a mixture of methane and carbon dioxide gas. Anaerobic digestion treats waste by converting putrid organic materials to carbon dioxide and methane gas. This gas is referred to as biogas.

The biogas can be used to produce both electrical power and heat. The conversion of solids to biogas results in a much smaller quantity of solids that must be disposed. During the anaerobic treatment process, organic nitrogen compounds are converted to ammonia, sulfur compounds are converted to hydrogen sulfide, phosphorus to orthophosphates, and calcium, magnesium, and sodium are converted to a variety of salts. Through proper operation, the inorganic constituents can be converted to a variety of beneficial products. The end products of anaerobic digestion are natural gas (methane) for energy production, heat produced from energy production, a nutrient rich organic slurry, and other marketable inorganic products.

The effluent containing particulate and soluble organic and inorganic materials can be separated into its particulate and soluble constituents. The particulate solids can be sold or exported from the dairy while the nutrient rich liquids are applied to the land.

Bacterial Consortia

Anaerobic digestion is carried out by a group, or consortia of bacteria, working together to convert organic matter to gas and inorganic constituents. The first step of anaerobic digestion is the breakdown of particulate matter to soluble organic constituents that can be processed through the bacterial cell wall. Hydrolysis, or the liquification of insoluble materials is the rate-limiting step in anaerobic digestion of waste slurries.

This step is carried out by a variety of bacteria through the release of extra-cellular enzymes that reside in close proximity to the bacteria. The soluble organic materials that are produced through hydrolysis consist of sugars, fatty acids, and amino acids. Those soluble constituents are converted to carbon dioxide and a variety of short chain organic acids by acid forming bacteria. Other groups of bacteria reduce the hydrogen toxicity by scavenging hydrogen to produce ammonia, hydrogen sulfide, and methane.

A group of methanogens converts acetic acid to methane gas. A wide variety of physical, chemical, and biological reactions take place. The bacterial consortia catalyze these reactions. Consequently, the most important factor in converting waste to gas is the bacterial consortia. The bacterial consortia are essentially the "bio-enzymes" that accomplish the desired treatment. A poorly developed or stressed bacterial consortium will not provide the desired conversion of waste to gas and other beneficial products.

Factors Controlling the Conversion of Waste to Gas

The rate and efficiency of the anaerobic digestion process is controlled by:

- The type of waste being digested,
- Its concentration,
- Its temperature,
- The presence of toxic materials,
- The pH and alkalinity,
- The hydraulic retention time,
- The solids retention time,
- The ratio of food to microorganisms,
- The rate of digester loading,
- And the rate at which toxic end products of digestion are removed.

Waste Characteristics

All waste constituents are not equally degraded or converted to gas through anaerobic digestion. Anaerobic bacteria do not degrade lignin and some other hydrocarbons. The digestion of waste containing high nitrogen and sulfur concentrations can produce toxic concentrations of ammonia and hydrogen sulfide. Wastes that are not particularly water-soluble will breakdown slowly. Dairy wastes have been reported to degrade slower than swine or poultry manure.

As pointed out earlier, lignin will not degrade during anaerobic digestion. Since a substantial portion of the volatile solids in dairy waste is lignin, the percentage of cow manure volatile solids that can be converted to gas is lower when compared to other manure and wastes. The manure characteristics also establish the percentage of carbon dioxide and methane in the biogas produced. Dairy waste biogas will typically be composed of 55 to 65% methane and 35 to 45% carbon dioxide. Trace quantities of hydrogen sulfide and nitrogen will also be present.

Dilution of Waste

The waste characteristics can be altered by simple dilution. Water will reduce the concentration of certain constituents such as nitrogen and sulfur that produce products (ammonia and hydrogen sulfide) that are inhibitory to the anaerobic digestion process. High solids digestion creates high concentrations of end products that inhibit anaerobic decomposition. Therefore, some dilution can have positive effects.

The literature indicates that greater reduction efficiencies occur at concentrations of approximately 6 to 7 percent total solids. Dairy waste "as excreted" is approximately 12 percent total solids and 10.5 percent volatile solids. Most treatment systems operate at a lower solids concentration than the "as excreted" values.

Component	*% Dry of Matter*
Volatile solids	83.0
Ether Extract	2.6
Cellulose	31.0
Hemicellulose	12.0
Lignin	12.2
Starch	12.5
Crude Protein	12.5
Ammonia	0.5
Acids	0.1

Dilution also causes stratification within the digester. Undigested straw forms a thick mat on top of the digester while sand accumulates at the bottom. The optimum waste concentration is based on temperature and the quantity of straw and other constituents that are likely to separate within the anaerobic digester. It is desirable to keep the separation or stratification in the digester to a minimum. Intense mixing involving the consumption of power may reduce the stratification of dilute waste.

The use of flush systems to remove the manure from the dairy barns has major economic advantages to the dairy. Flush systems normally use 100 to 200 gallons per cow, per day of dilution water. The flush volumes required are based on the lane or gutter length, width and slope (Fulhage and Martin 1994). The flush water usually contains very low concentrations of total and volatile solids. At 100 gallons per cow of flush water, the waste has only 12.5 percent of the "as excreted" concentration. At 200 gallons per cow per day of flush water the waste contains only 6.25 percent of the "as excreted" concentration. The milk parlor also produces a substantial amount of dilute waste. Approximately 15% of the animal manure is deposited in the milk parlor.

Foreign Materials

Addition of foreign materials such as animal bedding, sand and silt can have a significant impact on the anaerobic digestion process. For example, the poor performance of the Monroe, WA dairy digester was attributed to the use of cedar wood chip bedding (Ecotope 1979). The quantity and quality of the bedding material added to the manure will have a significant impact on the anaerobic digestion of dairy waste. Sand and silt must be removed before anaerobic digestion. If it is not removed before digestion it must be suspended during the digestion process.

Toxic Materials

Toxic materials such as fungicides and antibacterial agents can have an adverse effect on anaerobic digestion. The anaerobic process can handle small quantities of toxic materials without difficulty. Storage containers for fungicides and antibacterial agents should be placed at locations that will not discharge to the anaerobic digester.

Nutrients

Bacteria require a sufficient concentration of nutrients to achieve optimum growth. The carbon to nitrogen ratio in the waste should be less than 43. The carbon to phosphorus ratio should be less than 187. Hills and Roberts showed that a non-lignin C/N ratio of 20 to 25 is optimum for digester performance. Typically, "as excreted manure has a C/N ratio of 10.

Temperature

The anaerobic bacterial consortia function under three temperature ranges.

Psychrophilic temperatures of less than 68 degrees Fahrenheit produce the least amount of bacterial action. Mesophilic digestion occurs between 68 degrees and 105 degrees Fahrenheit. Thermophilic digestion occurs between 110 degrees Fahrenheit and 160 degrees Fahrenheit. The optimum mesophilic temperature is between 95 and 98 degrees Fahrenheit.

The optimum thermophilic temperature is between 140 and 145 degrees Fahrenheit. The rate of bacterial growth and waste degradation is faster under thermophilic conditions. On the other hand, thermophilic digestion produces an odorous effluent when compared to mesophilic digestion. Thermophilic digestion substantially increases the heat energy required for the process. In most cases, sufficient heat is not available to operate in the thermophilic range. This is especially true if flush systems are used or the milk parlor waste is mixed with the scraped manure. Large quantities of dilution flush water must be heated to the digester's operating temperature.

During cold weather, control of the flush volume is critical in maintaining adequate digester temperatures. Seasonal and diurnal temperature fluctuations significantly affect anaerobic digestion and the quantities of gas produced. Bacterial storage and operational controls must be incorporated in the process design to maintain process stability under a variety of temperature conditions.

Temperature is a universal process variable. It influences the rate of bacterial action as well as the quantity of moisture in the biogas. The biogas moisture content increases exponentially with temperature. Temperature also influences the quantity of gas and volatile organic substances dissolved in solution as well as the concentration of ammonia and hydrogen sulfide gas.

pH

Methane producing bacteria require a neutral to slightly alkaline environment (pH 6.8 to 8.5) in order to produce methane. Acid forming bacteria grow much faster than methane forming bacteria. If acid-producing bacteria grow too fast, they may produce more acid than the methane forming bacteria can consume. Excess acid builds up in the system. The pH drops, and the system may become unbalanced, inhibiting the activity of methane forming bacteria. Methane production may stop entirely. Maintenance of a large active quantity of methane producing bacteria prevents pH instability. Retained biomass systems are inherently more stable than bacterial growth based systems such as completely mixed and plug flow digesters.

Hydraulic Retention Time (HRT)

Most anaerobic systems are designed to retain the waste for a fixed number of days. The number of days the materials stays in the tank is called the Hydraulic Retention Time or HRT. The Hydraulic Retention Time equals the volume of the tank divided by the daily flow (HRT=V/Q). The hydraulic retention time is important since it establishes the quantity of time available for bacterial growth and subsequent conversion of the organic material to gas. A direct relationship exists between the hydraulic retention time and the volatile solids converted to gas.

Solids Retention Time (SRT)

The Solids Retention Time (SRT) is the most important factor controlling the conversion of solids to gas. It is also the most important factor in maintaining digester stability. Although the calculation of the solids retention time is often improperly stated, it is the quantity of solids maintained in the digester divided by the quantity of solids wasted each day.

Where V is the digester volume; Cd is the solids concentration in the digester; Qw is the volume wasted each day and Cw is the solids concentration of the waste. In a conventional completely mixed, or plug flow digester, the HRT equals the SRT. However, in a variety of retained biomass reactors the SRT exceeds the HRT. As a result, the retained biomass digesters can be much smaller while achieving the same solids conversion to gas.

The volatile solids conversion to gas is a function of SRT (Solids Retention Time) rather than HRT. At a low SRT sufficient time is not available for the bacteria to grow and replace the bacteria lost in the effluent. If the rate of bacterial loss exceeds the rate of bacteria growth, “wash-out” occurs. The SRT at which “wash-out” begins to occur is the “critical SRT”.

Jewel established that a maximum of 65 percent of dairy manure’s volatile solids could be converted to gas with long solids retention times. Burke established that 65 to 67 percent of dairy manure COD could be converted to gas. Long retention times are required for the conversion of cellulose to gas.

The goal of process engineers over the past twenty years has been to develop anaerobic processes that retain biomass in a variety of forms such that the SRT can be increased while the HRT is decreased. The goal has been to retain, rather than waste the biocatalyst (bacterial consortia) responsible for the anaerobic process. As a result of this effort, gas yields have increased and digester volumes decreased. A measure of the success of biomass retention is the SRT/HRT ratio. In conventional digesters, the ratio is 1.0. Effective retention systems will have SRT/HRT ratios exceeding 3.0. At an SRT/HRT ratio of 3.0 the digester will be 1/3rd the size of a conventional digester.

Digester Loading (kg / m3 / d)

Neither the hydraulic retention time (HRT), nor the solids retention time (SRT) tells the full story of the impact that the influent waste concentration has on the anaerobic digester. One waste may be dilute and the other concentrated. The concentrated waste will produce more gas per gallon and affect the digester to a much greater extent than the diluted waste. A more appropriate measure of the waste on the digester's size and performance is the loading. The loading can be reported in pounds of waste (influent concentration x influent flow) per cubic foot of digester volume. The more common units are kilograms of influent waste per cubic meter of digester volume per day (kg / m3 / d). One (kg / m3 / d) is equal to 0.0624 (lb / ft3 / d). The digester loading can be calculated if the HRT and influent waste concentration are known.

Food to Microorganism Ratio

The food to microorganism ratio is *the key factor* controlling anaerobic digestion. At a given temperature, the bacterial consortia can only consume a limited amount of food each day. In order to consume the required number of pounds of waste one must supply the proper number of pounds of bacteria. The ratio of the pounds of waste supplied to the pounds of bacteria available to consume the waste is the food to microorganism ratio (F/M). This ratio is the controlling factor in all biological treatment processes. A lower the F/M ratio will result in a greater percentage of the waste being converted to gas.

Unfortunately, the bacterial mass is difficult to measure since it is difficult to differentiate the bacterial mass from the influent waste. The task would be easier if *all* of the influent waste were converted to biomass or gas. In that case, the F/M ratio would simply be the digester loading divided by the concentration of volatile solids (biomass) in the digester (L / Cd). For any given loading, the efficiency can be improved by lowering the F/M ratio by *increasing the concentration of biomass in the digester*. Also for any given biomass concentration within the digester, the efficiency can be improved by decreasing the loading. Unfortunately, a portion of the influent waste is not processed or converted to biomass or gas by the bacteria. In that case the F/M ratio is equal to the VS loading divided by the digester VS measured (VSD) minus the unprocessed Volatile Solids (VSUP). The unprocessed volatile solids may include refractory or non-degradable biological products produced by the bacteria.

End Product Removal

The end products of anaerobic digestion can adversely affect the digestion process. Such products of anaerobic digestion include organic acids, ammonia nitrogen, and hydrogen sulfide. For any given volatile solids conversion to gas, the higher the influent waste concentration, the greater the end product concentration. End product inhibition can be reduced by lowering the influent waste concentration or by separately removing the soluble end products from the digester through elutriation. Elutriation is the process of washing the solids (bacteria) with clean water to remove the products of digestion. The contact process provides an efficient means of removing the end products of digestion. End product removal can be enhanced by elutriation, which is easily incorporated into the contact process (Burke 1997).

Digester Types

A vast array of anaerobic digesters have been developed and placed in operation over the past fifty years. A variety of schemes could be used to classify the digestion processes. For dairy waste, the most important classification is whether or not it can be used to convert dairy waste solids to gas while meeting the goals of anaerobic digestion. The goals of dairy waste anaerobic digestion are as follows:

1. Reduce the mass of solids
2. Reduce the odors associated with the waste products
3. Produce clean effluent for recycle and irrigation
4. Concentrate the nutrients in a solid product for storage or export
5. Generate energy
6. Reduce pathogens associated with the waste.

In addition, the digester must be able to handle or process the dairy waste stream. Dairy waste is a semi-solid slurry. Much of the energy value is in the solids. Consequently, the process must be able to convert solids to gas without clogging the anaerobic reactor. The process must also be able to handle bedding material, sand and other foreign materials associated with typical dairy waste. In addition, if the dairy manure is a dilute waste, the process must be capable of mitigating stratification and solids separation within the reactor.

Processes that are not Appropriate for Digesting Dairy Manure

A variety of high rate anaerobic processes, which retain bacteria have been developed to treat soluble organic industrial wastes. These "high rate" digesters have reduced hydraulic detention times from 20 days to a few hours. They include anaerobic filters, both upflow and downflow, and a variety of biofilm processes such as fixed film packed bed reactors. Bacteria are retained in these reactors as films on carriers such as plastic beads, or sand, or on support media of all configurations. The waste washes past the retained bacteria. The bacteria convert the soluble constituents to gas but have little opportunity to hydrolyze and degrade the particulate solids, unless the solids become attached to the biomass.

These reactors are *not* suitable for digesting dairy waste since they are not effective in converting particulate solids to gas and tend to clog while digesting dairy manure slurries. These high rate reactors can treat the soluble component of dairy waste. But only a fraction of the available energy will be recovered.

A widely used industrial waste anaerobic digester is the UASB or "Upflow Anaerobic Sludge Blanket", reactor. The process stores the anaerobic consortia as pellets, approximately the size of a pea. The upflow anaerobic sludge blanket reactor (UASB) is widely used in industrial treatment processes throughout the world. It is an extremely effective process for converting soluble organic materials, such as sugar to methane gas. It has not been used for processing dairy waste since it is ineffective in converting solids to gas. It is primarily used to convert non-particulate or soluble waste to gas.

UASB Reactor

The anaerobic baffled reactor is a horizontal version of the upflow anaerobic sludge blanket reactor. Both store large quantities of anaerobic bacteria as pellets approximately the size of a pea. Unfortunately, these very successful anaerobic reactors are not effective in digesting particulate waste.

Particulate solids tend to settle in the horizontal baffled reactor (HBR) while organic fibers will form a mat on the surface. There are no known instances of the HBR being used for the treatment of dairy waste. Unless the dairy waste was thoroughly screened and all particulate matter removed the HBR would tend to become clogged. The removal of solids by screening and gravity sedimentation will eliminate up to 80% of the energy generating potential from dairy waste.

Processes that can be used for Digesting Dairy Manure

The processes that have been used for digesting dairy waste can be subdivided into high rate and low rate processes. Low rate processes consist of covered anaerobic lagoons, plug flow digesters, and mesophilic completely mixed digesters. High rate reactors include the thermophilic completely mixed digesters, anaerobic contact digesters, and hybrid contact/fixed film reactors.

Anaerobic Lagoons (Very Low Rate)

Anaerobic lagoons are covered ponds. Manure enters at one end and the effluent is removed at the other. The lagoons operate at psychrophilic, or ground temperatures. Consequently, the reaction rate is affected by seasonal variations in temperature.

Best Management Practices to Handle Dairy Wastes

Dairy farmers recognize the importance of water quality; they accept the responsibility of managing their dairy wastes to protect, preserve, and even improve the quality of both surface and ground waters. No single dairy waste management practice can meet the needs of every dairy nor is a single waste management practice appropriate or practical on all dairies in Alabama.

Wise dairy operators realize that the best answer to their situation will be a combination of waste management practices. Such dairy operators can seek input from several sources. Extension dairy production and waste management specialists and waste management and water quality specialists from NRCS and other agricultural agencies can help operators establish the best waste management system for their dairy.

This publication briefly describes various animal waste management practices. When properly applied to specific dairy situations, these activities become "best management practices" that help dairy operators:

- Prevent direct discharge of manure or wastewater into surface waters or onto adjacent neighbours' property.
- Prevent any nuisance conditions that interfere with normal use and enjoyment of neighbours' property.

- Enhance the operational efficiency of the dairy unit.
- Collect and use dairy manure and wastewater for beneficial purposes such as fertilizer, compost, or bedding.

Dairy operators in Alabama can use one or more of the following dairy waste management techniques to reach the goals stated above.

Holding Pond for Milking Centre Wastes

Dairy wastes associated with the milking centre are liquid and are easily collected in a holding pond close to the milking centre. This holding pond can also collect liquid wastes and runoff from holding areas. The pond should be sized to hold this liquid waste plus any rainfall that occurs between pond pump-out intervals.

Runoff Management

Noncattle open areas around the dairy milking centre or dairy barn facility should be sloped to direct freshwater runoff away from cattle areas and manure collection structures unless additional water is needed for irrigation. Unless sized for this additional runoff, ponds and lagoons will fill and require pumping out quicker than expected. For wastewater irrigation, all runoff can be directed into either the milking centre holding pond, waste storage pond, or lagoon system. This allows for land application with other collected dairy waste. Roof guttering is one way of collecting roof runoff and controlling where it goes, either away from or into the waste stream as desired.

Scrape and Haul

This system uses a scraping operation to remove solid manure from holding areas and even some freestall barns to a special manure holding area to await hauling for land application. This special manure holding area should have an impervious base such as compacted clay, but concrete is preferable. This impervious base should be sloped to drain liquid manure away into a waste storage pond that also handles milking centre liquid wastes. Storage areas may also be covered to reduce rainfall runoff into the system. This method of dairy waste management is very basic and is generally most applicable for dairy herds with less than 100 cows. In some instances, the manure hauling and land application interval could be daily, but more than likely it will be weekly or monthly.

Waste Storage Pond

A dairy waste storage pond is a specially constructed pond used to collect and store manure, flush water, and polluted runoff from a dairy facility. Storage is a relatively short period of 90 to 180 days. Waste storage pond contents must be removed at the end of this storage period with land application the most common end use. The waste storage pond must be constructed with an impermeable liner to prevent waste leakage to ground water. Because waste storage ponds produce an odour, they should be located downwind of neighbours, highways, and any public use area. Storage pond waste retains most of its fertilizer nutrients. As a result, at a land application rate of 200 pounds of nitrogen per acre, more than two and one-half times as much cropland will be needed for land application of waste from a storage pond system than from a two-stage lagoon system for the same number of dairy cows.

Dairy Waste Lagoon

Dairy waste lagoons are earthen structures designed for biological treatment and long-term storage of dairy waste. Lagoons are specially constructed to prevent leakage of dairy waste to ground water. The lagoon system allows manure to be handled with water-flushing systems, sewer lines, pumps, and irrigation equipment. The natural biological action on the waste results in less odour during land application. Nitrogen content of the waste is reduced in lagoons by as much as 80 percent.

This reduction minimizes land area needed for land application and enhances long-term storage.

Waste Storage Pond Agitation and Lagoon Renovation

Agitation prior to and during both waste storage pond pumpout and lagoon renovation is necessary to suspend waste solids that have settled to the bottom or floated to the surface.

This agitation will be by either a portable propeller-type pump or a chopper-type pump requiring at least a 100-horsepower tractor. Agitation should thoroughly mix the solid and liquid contents to allow removal of both during the pump-out process.

Solid/Liquid Separation for Dairy Waste

Twelve to 14 percent of excreted dairy manure is solids. This, along with any solid bedding material used in freestall housing, will reduce storage pond or lagoon capacity. Even after biological treatment, the lagoon will be quickly filled. These accumulated solids also greatly reduce lagoon efficiency. Solid-liquid separation using either settling basins or mechanical separators (stationary screens or elevators) can reduce up to half of the solids in the waste stream and should be considered even when no bedding is used. These solids can then be used for fertilizer, bedding, feed, or composting. If used, separators can increase the capacity of storage ponds and waste lagoons.

Land Application

Land application is the best end use of dairy manure collected regardless of the handling method. Solids are applied with manure spreaders, and liquid contents of storage ponds and dairy lagoons are land applied through irrigation.

Land application should match fertilizer requirements of the target crop with plant-available nutrients in the waste. Matching crop fertilizer requirements minimizes harmful effects to water quality that can occur through runoff into surface water or deep percolation into ground water. It also reduces the risk of water pollution from manure application to the same level as manufactured fertilizer application. Timing of applications and soil incorporation should also be considered.

Waste Testing

Proper land application of waste requires nutrient content of the waste to be analyzed or estimated. While laboratory analysis of lagoon wastewater and agitated lagoon slurry is available, a 3- to 10-day time lag between sampling and receiving analysis results is typical. Field tests for plant-available nitrogen are available to allow on-site wastewater and dairy slurry testing during the land application process. This allows calibration of the land application process to provide the proper rate of nutrient application for a crop.

Soil Testing

Prior to land application of dairy wastewater or slurry waste, the target field intended for waste application should be soil tested. Soil testing tells dairy operators how much fertilizer is required for the crop and allows them to match the dairy waste/wastewater application to meet these crop fertilizer requirements.

Wastewater Irrigation

Although tractor-pulled tanks with injection of wastes are available, they are not practical for dairies in Alabama.

Using regular agricultural irrigation pumps and either travelling guns or solid set or centre pivot irrigation equipment to land apply dairy wastewater (liquid with less than 2 percent solids content) is the best way to handle large volumes of liquid wastes generated by a freestall flush system. Irrigation equipment should be calibrated to match crop fertilizer requirements.

Stream Exclusion

In those pasture areas where dairy animals have access to flowing streams, steps should be taken to minimize cattle loitering in the stream. Heavy or high-use access could cause stream bank erosion or destruction of stream side vegetation and water quality reduction due to manure deposits. Stream shade removal, fences, and other barriers may be used to protect the quality of streams flowing through these pastures.

Rotational Grazing

Dividing pastureland, either rainfed or irrigated, into small areas for controlled daily grazing allows opportunity for greater per feed unit milk production through intensive pasture management for grazing. Direct land application of dairy waste and wastewater promotes grass production and easier manure handling.

9

Bacteriological Identification and Antibiotic Sensitivity Testing

Clostridia are Gram-Positive rod-shaped spore-forming anaerobic bacteria. Their in vitro growth, identification in the routine diagnostic laboratory and antibiotic sensitivity testing must therefore follow specific rules and use specific materials. In this chapter the general diagnostic methodology, the specific growth requirements, the different orientation and identification assays and the antibiotic sensitivity testing of clostridia are described. In some of the sections, special emphasis will be placed on the growth and identification of Clostridium dificile. 13.1. General methodology Probably the most important prerequisite in the laboratory diagnosis of any clostridial species is a correct clinical sample sent following the strict requirements of anaerobes. Although the clostridial species differ in their requirements for anaerobiosis, with some being even aero-tolerant, all samples should be plated onto appropriate blood agar plates for anaerobic incubation and onto classical blood agar plates and selective agar plates for enterobacteria for comparative aerobic growth. During the incubation period of up to five days the plates should be examined daily for colony growth.

The whole procedure may sometimes take several days and is often not compatible with the field practise, in either human or veterinary medicine. But following the whole procedure is absolutely necessary to prevent the diagnostic laboratory from using identification and antibiotic sensitivity tests for anaerobes on non-strictly anaerobic bacteria. Precise clinical data from the doctor or the veterinarian are also of importance in the identification of the major and minor pathogen clostridia. Clostridia grow best on enriched media, but apart from being anaerobes, the majority of species have no particular growth requirements (with the exception of the intracellular C. piliforme, of course!).

Clostridia grow on commercially available 5% sheep blood agar plates for anaerobes, such as CDC anaerobe blood agar and Schaedler blood agar. Other classical 5% sheep blood agar, such as Columbia, Brucella and Brain-Heart Infusion, must be supplemented with yeast extract, vitamin K1 and haemin to favour growth of anaerobes (Bisping and Amtsberg, 1988; Quinn et al., 1994; Atlas and Park, 2000). Anaerobic storage or pre-incubation of the plates for a few hours in anaerobiosis enhances the chance of isolation of clostridia. Colony swarming can be prevented by growing the bacteria on phenylethyl alcohol blood agar (PEA) or on 4% agar media. The species of the genus Clostridium can also grow on dozens of selective blood agars for use in medical or food microbiology, several of which are commercially available, with, for instance, aminoglycosides or colistin/ nalidixic acid to inhibit coliforms and other Gram negative bacteria, or Columbia with cycloserin for isolation of C. perfringens (Manteca et al., 2001), or the selective media for C. dificile.One of the problems is the variation in the colony morphology and in the zone(s) of haemolysis when the clostridia grow on different media, leading to confusion in the pre-identification criteria. The origin of the blood is also of importance. The zone(s) of haemolysis by bovine isolates of C. perfringens toxin type A are broader and clearer when grown on ox blood agar than on sheep blood agar (Manteca, Mainil and Daube, unpublished data).

Growth is also possible in liquid media. The broth medium must be supplemented with serum and reducing agents, such as thioglycollate or cooked meat broths. An alternative is to add liver pieces (from rabbits) to BHI broth. The broth with a piece of rabbit liver is most probably the best liquid medium (Kaeckenbeeck and Mainil, unpublished), but its preparation is time-and labour-consuming. In most liquid media the clostridia grow best in the reduced portion of the medium, i.e. at the bottom of the tube, close to the piece of meat or of liver. All liquid media must be heated and boiled for 5-10 minutes prior to inoculation, in order to expel the absorbed traces of O_2, followed by cooling at 37°C (Quinn et al., 1994).

Because they are anaerobes, the clostridia were separated early from the genus Bacillus (Mainil et al., 2004). In addition, all clostridia prefer to grow in an atmosphere with 2 to 10% CO_2 (Quinn et al., 1994). Nevertheless, variations exist in their anaerobic requirements. Some species are very strict anaerobes and do not grow in presence of even traces of O_2. Conversely, other species are aerotolerant (C. histolyticum, C. tertium) and can still grow either in a mixture of 5% O_2, 10% CO_2, 85% N_2, or even in air, though less extensively than in absence of O_2. In contrast with Bacillus species, Clostridium species can sporulate only in absence of O_2 (Peck et al., 2004).

Colony growth normally takes 24h to 48h at a near-neutral pH and 37°C, but may be delayed up to 5 days for some species (C. botulinum, C. colinum) or in nonoptimal conditions. Incubation inside an anaerobic cabinet is the best alternative and is very flexible, allowing daily examination of the plates. But this equipment is of course very expensive. Incubation in anaerobic jars with envelope systems (GasPak) generating anaerobiosis (H_2 + CO_2) is the second best alternative.

However daily examination becomes costly since new packs must be used each time. Moreover, contact with O_2, even shortly, can already reduce the viability of the most strictly anaerobic species. The third alternative is the anaerobic jar with replacement of the air by N_2 and 5-10% CO_2. A catalyst must be present inside the jars to remove the remaining O_2 traces and to allow the growth of strict anaerobes.

However, most often this system does not favour the growth of the most strictly anaerobic clostridia, while favouring the growth of more aerotolerant species, leading to diagnostic confusions and errors.

Liquid media have the advantage of being aerobically incubated. However, one must also be careful, since not all of them support the growth of the most strictly anaerobic species. Whatever system is used, a strictly aerobic bacterial species, like Pseudomonas aeruginosa, and a strictly anaerobic bacterial species should be grown in parallel as controls for anaerobiosis.

Just after the discovery of the pathogenic role of C. dificile, George et al. proposed a selective agar plate called CCFA (cycloserine cefoxitin fructose agar) (George et al., 1979) for isolation from stool specimens of humans, horses and pet animals. This medium containing yeast extract, haemin, vitamin K, and cysteine is still used in most laboratories. The selective agents are cycloserine at the concentration of 500 mg/L and cefoxitine at 16 mg/L. These concentrations have been reduced to 250 mg/L and 10 mg/L respectively in some studies (Levett,

1985). Cefotaxime may also be used instead of cefoxitine at the concentration of 4 mg/L. The original formulation included egg-yolk, which may be replaced by blood. This is commercially available from several companies. Stools are directly inoculated and incubated in an anaerobic atmosphere for 48h. Anaerobic pre-incubation of the plate may also improve the recovery rate (Peterson et al., 1996).

The use of an Anoxomat system, which enables the obtention of an anaerobic atmosphere in the jar within one minute, allows the reading of the plates after only a 24 hour incubation period (Delmée, unpublished data). It is worth saying that, in the vast majority of cases, the use of a standard CCFA medium with a standard inoculation procedure is quite satisfactory for the diagnosis of most cases of C. dificile-associated diarrhoea. Culture is the most sensitive method but it is not very specific, due to the possibility of isolating nontoxigenic isolates. It is a slow method, but one that allows the testing of strains for toxigenicity. It is also the only method for epidemiological investigations.

The shapes of the colony and of the haemolysis zone(s) are important orientation criteria. Gram and Ziehl-Neelsen (acid-fast) stained smears must be performed to study the morphology of the cells and of the spores. Any suspect colony must subsequently be subcultured anaerobically, aerobically and microaerobically to prove the requirement for anaerobiosis, before applying the identification tests (without forgetting that a few minor pathogen species are aerotolerant). Identification is performed by study of the biochemical properties, by gas liquid chromatography and by molecular methods, including toxin typing. In strict anaerobic conditions, colonies appear within 24 to 48 hours. Some species are, however, fastidious growers and require longer incubation. The three criteria to examine carefully are the colony morphology (surface, form), the presence/absence of colony swarming and the haemolysis pattern (number and width of haemolysis zones).

The orientation tests are generally similar to those for non anaerobes : Gram and Ziehl-Neelsen stained smears, catalase, cytochrome peroxidase and motility tests. Production of lecithinase and lipase are also considered useful preidentification tests. Though clostridia are Gram-Positive bacteria, Gram stained smears display a mixture of Gram-Positive and Gram-Negative bacterial cells after in vitro growth. In Gram stained smears, most clostridia appear as straight or curved rods, but the shape of a few species can also vary from coccoid rods to long filamentous forms.

The ends of the rods are rounded, tapped or blunt. The cells of many species occur singly or in pairs and several can also form chains of varying lengths and numbers of cells. In chains, the individual cells are of similar length (C. chauvoei) or can be of greatly different lengths (C. septicum). The chains formed by a few species can form tight coils or spiral-shaped (C. spiroforme) configurations. Specific arrangements are, however, more frequent and typical in smears from the lesions than in smears from the colonies. The shape and arrangements of the major pathogen clostridia in Gram stained smears from colonies are of course very important for preidentification, but other minor or non pathogen clostridial species can resemble one or other of the major pathogen species.

In modified Ziehl-Neelsen (acid-fast) stained smears, spores are ovoid or spherical, centrally, subterminally or terminally localized, and cell-distending. Some species sporulate very easily and quickly, while others only rarely. Clostridia give negative results at the catalase (in contrast with Bacillus species) and cytochrome peroxidase tests. Some species can, however, give a weakly positive result at the catalase test. On the other hand, most species are motile, but quickly loose their motility in the presence of O_2.

Most clostridia are saccharolytic or proteolytic, or both, but a few species are neither. The biochemical-based minisystems are best suited for identification of saccharolytic, fast growing clostridia. With these systems the identification can be obtained after 24 to 48hrs of anaerobic incubation by reading the colour reactions manually. A code number is generated and interpreted from the code book or computerized database provided by the manufacturer. The rapid pre-existing enzyme-based minisystems utilise chromogenic or fluorogenic/chromogenic substrates for detection of bacterial glycosidases, aminopetidases and some other enzymes. They are best suited for rapid identification of slow growing clostridia. The inoculation of a heavy bacterial suspension, equal to McFarland standard n° 3 to 5, allows the reading of the results of the reactions after 4 hours of aerobic incubation. The identification can be reached either manually by using a code book or by a computer-assisted automated reader, such as ATB Expression. Nevertheless, none of the identification kits is 100% reliable, even for the identification of the major and minor pathogenic species (60-80%). So, final identification is completed by anti-whole cell antibody immunofluorescent assays (C. chauvoei, C. septicum, C. sordellii, C. novyi, C. haemolyticum), genetic tests, gas liquid chromatography and/or toxin typing assays, if necessary. Colonies of C. dificile are easily recognised on culture plates, due to their typical ground glass appearance when observed under a magnifying glass.A yellow-green or chartreuse fluorescence under ultraviolet illumination is another characteristic of the colonies, but this may vary with the medium used. The typical odour (horse manure) is also an aid to identification. Gas liquid chromatography (GLC) of an agar plot around a suspected colony is the best and easiest method for definite identification. C. dificile displays a typical GLC profile, with large amounts of butyric and iso-caproic acids. Such a device, however, is not available in many laboratories. Individual biochemical tests or anaerobic panels may also be used (rapid pre-existing enzyme-based minisystems are strongly recommended), as well as a somatic antigen latex kit. Cross-reactions, however, have been documented with this latter reagent.

Production of proline-aminopeptidase by a disc test has been recently proposed as a means of rapid identification of C. dificile, when used in conjunction with the typical morphology of the colonies (Delmée, 2001, 2004). A direct enzyme immunoassay (EIA) for toxin A may also be used but then, of course, it will only recognise toxigenic isolates. Enzyme immuno-assays detecting glutamatedehydrogenase (GDH), an enzyme specific to C. dificile, can also help to identify this species, in faeces and after in vitro growth. In combination, these two EIAs for toxin A and for GDH can identify and differentiate non-toxigenic and toxigenic isolates of C. dificile, either after in vitro growth or directly in faeces (TRIAGE C. dificile [Biosite, San Diego, CA]). All or most clostridia species are intrinsically resistant to aminoglycosides (like all anaerobes), sulfonamides (C. perfringens is one exception) and diaminopyrimidines. Acquired resistance to members of other families of antibiotics, especially to the macrolides, lincosamides, tetracyclines and phenicols, has progressively developed at different frequency rates in clostridia species of clinical importance. Several clostridia species can indeed harbour antibiotic resistance genetic determinants on transferable molecules (plasmids, transposons, conjugative transposons) and may serve as reservoir of antibiotic resistance genes for other clostridial species and even genders.

For instance, resistance of human isolates of C. dificile and of human and ruminant isolates of C. perfringens to macrolides and lincosamides is mediated by different erm (Erythromycin Resistance Methylase) genes, possibly originating from Enterococcus or Streptococcus, some of which are located on a conjugative transposon or on a plasmid. Similarly, resistance to tetracyclines is most frequently mediated by the tetA gene amongst human, ruminant and poultry C.

perfringens isolates, that can be located on a conjugative plasmid, and by the tetM gene amongst human C. dificile isolates, that can be located on a conjugative transposon which is similar to a conjugative transposon of Enterococcus faecalis. A third example is the resistance of C. perfringens and C. dificile to chloramphenicol, which is mediated by two different CAT (Chloramphenicol Acetyl Transferase) enzymes in C. perfringens, one of which is synthesised by transposon located genes (catP) and is closely related to the catD genes of C. dificile, located on a similar transposon. Different percentages of isolates of other species are also resistant to these three families of antibiotics, but the mechanisms for this are mostly unknown Acquired resistance to other antibiotics, like b lactams (mainly penicillins, more rarely cephalosporins), glycopeptides, fluoroquinolones and metronidazole have also been described. In most species the percentage of resistant isolates remain low. Nevertheless, a high percentage of resistant isolates is found in some minor and major pathogen clostridia. These resistances develop as a consequence of mutations in chromosomal genes or after acquisition of new genes (such as those coding for the production of b lactamase active on penicillins in C. butyricum, C. clostridioforme, C. innocuum and C. ramosum) or are mediated by still unidentified mechanisms.

The determination of the sensitivity/resistance patterns of a clostridial isolate may be important clinically in human medicine, but much less in veterinary medicine. In human hospital medicine, clostridia are tested for antibiotic sensitivity in particular cases, such as the occurrence of several cases of infection or the persistence of an infection in a patient despite instauration of treatment(s) based on the intrinsic sensitivity of the species isolated. Another exception, in veterinary medicine, is probably C. dificile-associated diarrhoea in horses and pet animals. The main problem is the availability of the results only after 3 to 5 days, so that a treatment must be empirically initiated by the clinician. Testing the antibiotic sensitivity patterns of clostridia is therefore important mainly for epidemiological purposes in both human and veterinary medicines, not only for local hospitals, but also because these patterns can vary according to geography following the epidemiology of the specific resistance genetic determinants.

The antibiotics to test are narrow and/or broad spectrum b lactams (penicillins, cephalosporins and carbapenems, with b lactamase inhibitors), glycopeptides (vancomycin, teicoplanin), macrolides (erythromycin), lincosamides (clindamycin), phenicols (chloramphenicol, florfenicol), tetracyclines, recent fluoroquinolones, and/or metronidazole (Prescott et al., 2000). Macrolides can not be tested in anaerobic conditions with CO_2 concentrations between 5 and 10%. It must also be remembered that chloramphenicol (but not florfenicol) was banned years ago for use in farm animals, and that metronidazole has been banned more recently in the European Union. Bacitracin has been used for years in poultry to prevent enterotoxaemia caused by C. perfringens, but is today also banned in the European Union. Nitrofuranes are recommended against C. colinum infections in birds, but are also not allowed in farm animals in the European Union.

A variety of methods is available for susceptibility testing of anaerobic isolates, although problems are encountered with all of them. Agar plate methods include agar dilution, Wadsworth modified agar dilution, spiral gradient endpoint (SGE [Spiral System Instruments, Bethesda, MD]), PDM Epsilometer (E-test [AB Biodisk, Solna, Sweden]) and disk diffusion. Broth methods include broth dilution, broth micro-dilution and broth disk-elution. Not all of these procedures are currently endorsed (e.g. disk diffusion and broth disk-elution) by the NCCLS (National Committee for Clinical Laboratory Standard, 2004) whose guidelines are followed by the majority of the commercially available systems.

Agar dilution, broth macro-and micro-dilution are the techniques suggested to be used in research studies to determine the minimal inhibitory concentrations (MIC) of different antibiotics against a clostridial isolate in anaerobic conditions. The systems used in routine diagnostic laboratories are generally based on broth microdilution methods. There are commercial sources for frozen trays (MSI/Micro-Media System [Cleveland, OH], MicroTech Medical System [Aurora, CO] and Innovative Diagnostic Systems [Atlanta, GA]) or lyophilized trays [Radiometer America/Sensititre, Westlake, OH]). The limited number of appropriate antianaerobic drugs in some of these trays is a disadvantage, but the laboratories can be supplied with custom-poured MIC panels by MicroTech Medical System and by Sensititre.

E-test is another agar-diffusion method used successfully for anaerobes. It is quite easy to perform and to read; it shows a high categorical agreement with the agar dilution reference method in research studies but it is costly. Therefore this method is particularly suited for studying individual isolates against few selected agents.

In human anaerobic bacteriology the MIC values are interpreted and the tested strains categorized (Susceptible, Intermediate,

Resistant) according to the breakpoint concentrations recommended either by NCCLS, or by the EUCAST (European Committee on Antimicrobials Susceptibility Testing, 2003), or by national committee guidelines (e.g. Comité de l' Antibiogramme de la Société française de Microbiologie [2005], Deutsches Institute für Normung [2004]) in an attempt to achieve a good predictive value on the basis of in vivo correlations.

Molecular Typing and Identification Methods

The main aim of this chapter is to discuss the molecular characterisation of clostridia. However, some preliminary remarks on molecular identification in general are necessary to provide a background.Classification, characterisation and identification of micro-organisms in general Classification is the theory and process of ordering organisms into recognisable groups. Different classification schemes may exist in parallel. However, today the initial step of classification is mainly based upon characterisation of gene coding sequences for RNA of the small subunit of ribosomes (named 16S rRNA gene in the following pages), the sequences being the basis for unravelling evolutionary lines of descent.

In the pre-molecular era, classification systems were based, for example, on selected cellular and metabolic properties (such as presence or absence of spores and of photosynthetic apparatus) or the relationship to oxygen. Moreover, at the dawn of bacterial systematics, classification was even based on shape, colour or motility. It is obvious that characterisation and classification are inter-dependent, as the type of classification defines the range of properties to be determined. Present day classification, following the polyphasic approach, is much more demanding than any other previous classification strategy. This is because a broad spectrum of properties needs to be determined at the genetic level (information carried directly on the genes, including non-coding regions) and at the epigenetic level (information carried on the products of the genes).

Bacterial systematics consists of a series of sequential steps. The first step is characterisation, i.e. obtaining data on the properties of an organism. These properties are then compared to those of described species, usually following a dichotomic key or focusing on salient characteristics. Results will show whether the isolate can be identified, i.e., affiliated to a described taxon. If identification is not possible, the isolate is likely to represent a novel taxon which may be described following more extensive characterisation. The extent to which characterisation needs to be performed depends upon the range of properties assigned to phylogenetic neighbouring taxa, to which the new taxon belongs. If sufficient properties have been determined allowing the organisms to be classified (i.e. placed into a systematic framework), a name is given to the appropriate rank of the classified organism. It is named as a new species either of an established genus or of a novel genus. It may be concluded that the purpose and efficiency of a taxonomy is tested by its use in identifying organisms (Stackebrandt et al., 1998).

A chapter on molecular identification methods should not give the impression that non-molecular approaches have lost their significance (even though the molecular approach may one day replace these more traditional methods). In fact, molecular identification constitutes only one, albeit large, segment of a broader spectrum of methods required unambiguously to affiliate isolates to taxa today. Nucleic acid based identification approaches also have obvious advantages over more traditional methods. These are: cost, time, reproducibility, reliable authentification, existence of cumulative databases and their universal application. Nucleic acid based identification approaches are especially advantageous when large strain numbers from the clinical environment need to be identified rapidly, some expressing the most potent toxins known.

Nucleic acid stretches suitable for the identification of members of a given phylogenetic level are molecules showing some of the same characteristics also demanded of suitable phylogenetic markers. The stretches should be present among all representatives of a given group of phylogenetically related organisms in order to enable their study. Since every individual sequence position can only carry information on a rather narrow range of evolutionary time, an increasing number of independently evolving positions or regions raises the number of phylogenetic levels that can be detected.

Depending on the method applied, a nucleic acid stretch should contain at least one region of conserved structure to allow one PCR primer. The conserved region would also have a second primer targeting a taxon-specific variable stretch, to amplify a specific DNA fragment of a wide range of taxa. Between these conserved regions, the nucleotide composition should contain one or more regions of high sequence divergence (e.g., for sequence analyses of intergenic space [ITS] regions and intrageneric stretches). In addition, the rrn operons (coding for rRNA) should contain sequence idiosyncrasies (to allow fluorescent in situ hybridisation [FISH]), together with discriminative sequence analyses and/or electrophoretic separation of DNA fragment of different lengths.

The sequence should contain one, preferably more, restriction sites to allow generation of DNA patterns by Amplified rRNA Digest Restriction Analyses (ARDRA), a 16S rRNA gene-based RFLP method. The availability of the rapidly increasing database of sequenced genes and of completely sequenced genomes will facilitate the comparative search for nucleic acids suitable for identification. Likewise, the following will facilitate the development of identification protocols: the curated database of certain housekeeping genes (MLST, Multi-Locus Sequence Typing), search programs for tandem repeats (MLVA, Multi-Locus Variable-number tandem repeat Analysis) and probe design (ARB software).

Determination of the 16S rRNA gene sequence is among the first steps applied in the identification process. Either purified nucleic acid preparations or crude extracts of bacterial cells can be used for in vitro amplification of rrn genes (= rDNA). In principle, a few bacterial cells are sufficient to perform the analysis. The worldwide application of rDNA gene sequencing makes description of a detailed protocol unnecessary. The reader is referred to the publication by Rainey and Stackebrandt (1992); PCR primer sequences have been indicated by Lane (1991) and updates are regularly included in recent publications on the phylogeny and systematics of prokaryotes. The almost complete gene is amplified using two primers located at the 5' terminus (position 27 forward) and the 3' terminus (position 1525 reverse). A series of internally binding primers (usually 14–20 bases in length) (Lane, 1991) allows rapid sequencing by means of the primer-directed chain termination method.

More than 98% of the Clostridium species type strains have been sequenced (Stackebrandt and Hippe, 2001), including pathogenic species (Stackebrandt et al.,121 1999), the remaining species being unavailable from public collections. Major findings of phylogenetic studies have demonstrated that the genus Clostridium constitutes a broad assembly of distantly related major branches within the Firmicutes (Collins et al., 1994). Studies have also shown that the species C. botulinum, phenotypically circumscribed by the presence of neurotoxins, embraces at least five moderately related strain clusters, each being defined by the presence of a unique neurotoxin. Nearly full length 16S rRNA gene sequences have been subjected to RFLP analysis, digesting the amplificate with restriction enzymes (e.g. TaqI, RsaI), resulting in the discrimination of C. perfringens from other species (Cordoba et al., 2001) and in the classification of pathogenic strains (Gürtler et al., 1991).

As the position of variable regions is known where the majority of speciesspecific nucleotide differences occur, only parts of the complete gene need to be sequenced. Within the 16S rRNA gene, variable regions are located between the following positions: (I) 60 and 106, (II) 146 and 219, (III) 440 and 494, (IV) 588 and 651, (V) 821 and 879, (VI) 997 and 1044, (VII) 1117 and 1154, (VIII) 1240 and 1296, as well as (IX) 1435 and 1466 (E. coli numbering). The degree of variation differs between members of the main lines of descent (phyla and orders).

Amplification of rDNA stretches by polymerase chain reaction (PCR), flanking the variable regions, provides easy access to sequenceable material. The sequences of the members of Clostridium differ mainly in the variable regions located in region I. Long insertions with moderate sequence similarities have been detected in variable region I in Clostridium paradoxum (Rainey et al., 1996) and in members of the genus Desulfotomaculum (Stackebrandt et al., 1997), a neighbouring taxon of Clostridium. The multiple rrn operons may show strong micro-heterogeneity, requiring a cloning step for unambiguous sequence determination.

Regions displaying length heterogeneity are ideal candidates for identification protocols based on electrophoretic analysis of PCR products, especially when species-specific PCR primers are used. Application of two species-specific PCR primers, followed by restriction analysis of one of the resulting fragments, has led to the discrimination of C. estertheticum from other organisms detected in vacuum-packed meat (Helps et al., 1999). Sensitivity can be increased by application of nested PCR, where amplification is achieved by applying two sets of PCR primers, the second one amplifying a subset of the fragments generated by the first set of primers (Klijn et al., 1995)

The identification of highly variable regions with the gene coding for 16S rRNA has led to the development of oligonucleotide probes directed against these mainly species-specific regions. The phylogenetically broader the taxonomic group, the more difficult it is to design a single probe capable of detecting all taxon members. This is especially true for members of the genus Clostridium, which is phylogenetically heterogeneous. A broad group would also detect members of other genera that cluster with certain Clostridium species (e.g. Ruminococcus, Eubacterium, Sarcina), while a species-cluster-specific probe would exclude detection of a major part of the Clostridium species.

Originally, radioactively and digoxigenine labelled probes (Rainey and Stackebrandt, 1992) were designed that hybridised with one of the following: membrane-bound rRNA, isolated DNA or DNA released from lysed filter-fixed colonies. Examples were C. perfringens-specific probes (Rönner et al., 1994a) and probes directed against C. botulinum. Later, fluorescently labelled probes were developed that were able to target RNA directly within the cells. Application of multiple probes carrying different fluorescent labels to the same 16S rRNA, increased sensitivity and allowed detection of non-cultured cells directly within the environment (Amann et al., 1995).

Little data is available for clostridial species, as detection of these organisms in environmental samples has not yet been of high research priority. Examples include the use of fluorescence in situ hybridisation (FISH) probes in the enumeration of human colon bacteria (Harmsen et al., 2000a) and intestinal flora in breast-fed and formula-fed newborn infants (Harmsen et al., 2000b). A combination of a method of PCR amplification of a 16S rRNA gene region, followed by probe hybridisation, increases the sensitivity of detection of Clostridium tyrobutyricum, a bacterium involved in the late blowing of certain cheeses (Klijn et al., 1995).

Some data is available for the 23S rRNA gene, though this molecule is mainly used for phylogenetic studies. Partial 23S rRNA gene sequences have been used to detect C. chauvoei in clinical materials (Sasaki et al., 2000). Using primers directed against the 3'terminus of 16S rDNA and the 5'terminus of 23S rDNA, the intergenic spacer region is accessible for amplification followed by electrophoretic separation of the resulting fragments. One example is the molecular differentiation of some clostridia associated with "blown pack" spoilage (Broda et al., 2003), a species causing the clinical diagnosis in cows of blackleg or malignant oedema. A further example is the identification of intestinal clostridia, including several species pathogenic to humans and animals (Song et al., 2002).

This method has been introduced to base population genetics and epidemiology on a molecular level. Stretches of about 400–500 nucleotides of seven genes, coding for housekeeping proteins, are carefully analysed and differences in alleles recorded. Examples of genes used in MLST are DMSO reductase, aspartokinase glutamin synthase, thymidilate kinase, phosophomannomutase and anthranilate synthase. The selection of genes needs to be evaluated for each new taxon undergoing investigation, as genes common to a particular set of organisms may not be universally distributed. Prominent examples are strain-rich species such as Neisseria gonorrhoea and N. meningitidis, Yersinia pestis and Y. pseudotuberculosis, as well as Bacillus anthracis and B. cereus. The evolutionary distance between strains is quantified in MLST as the number of differing loci. All strains identical to a particular strain at 5 or more loci are considered members of a clonal complex, or "ecotype". Each species consists of several ecotypes, which should be considered the starting point for independent evolution from within a species as defined today.

The topology of trees of housekeeping genes not involved in recombination is similar to that of trees representing the core of the genome, i.e. genes coding for rRNA. These organisms are considered clonal organisms. In cases where the topologies of these two sets of genes differ, recombination has occurred and these organisms are considered panmictic. Epidemic populations are panmictic, but contain islands of clonal structure, i.e., the population structure is not coherent. No MLST data is as yet available for clostridia, but will very likely be generated because of the availability of strain-rich species with significant pathogenic potential.

The analyses of DNA patterns comprise a broad range of different techniques involving electrophoreses, including restriction fragment analyses and the specific and non-specific generation of PCR-mediated amplification products. This method should be considered a traditional approach. Originally it was used in two-dimensional physical mapping by separation of macro-restriction fragments using pulsed field gel electrophoresis (PFGE). This included a wide spectrum of different variations, for example: CHEF (Contour-clamped homogeneous electric field), FIGE (Field inversion gel electrophoresis), TAFE (Transverse alternating field electrophoresis) and PHOGE (Pulsed homogeneous orthogonal field gel electrophoresis). The advantage of high one-dimensional electrophoretic RFLP resolution was soon evaluated for taxonomic purposes and is used today for the discrimination of highly related strains. The quality of results depends on the generation of high molecular DNA and reproducible separation techniques, as well as on sophisticated software programs to recognise small differences in migration behaviour of fragments. Only a small amount of data is available for members of Clostridium, such as in the study by Cordoba et al. (2001).

This study differentiated C. perfringens from other strains of other Clostridium species. AFLP is the method indexing variations in the whole genome, depending on the presence of restriction sites, and is one which is highly reproducible. The method consists of three steps. Firstly, total DNA is digested using 2 restriction enzymes and the resulting restriction fragments subsequently ligated to halfsite-specific adaptors. Secondly, a subset of the fragments is selectively amplified, using 2 restrictionsite/adaptor-specific primers. Use of fluorescently labelled primers (FAFLP), will refine this method. Thirdly, the PCR products (30–50 fragments of size 80–550 bp) are electrophoretically separated on polyacrylamide gels and fragments numerically evaluated by computer assistant analysis. The AFLP method has been used to detect C. novyi and C. perfringens in drug users (McLauchlin et al., 2002) and to investigate C. perfringens food poisoning in seven outbreaks in the UK. A comparison of PFGE and AFLP has been published by Autio et al. (2003) concerning persistent and sporadic Listeria monocytogenes strains, distant relatives of Clostridium genus members.

The most recent development using DNA patterns for characterisation, identification and epidemiology uses the polymorphism of tandem repeats (TR) as informative markers. Although this technique has not yet been used for differentiation of clostridial strains, it should be briefly mentioned here, as its application to clostridia can be anticipated. Originally found in mammalian cells, TR reminiscent of mini-and micro-satellites have also been found in pathogenic prokaryotes (van Belkum et al., 1997). This is despite the fact that these species constitute genetically very homogeneous entities. The role of TR in the regulation and translation of genes has been discussed by Le Flche et al. (2001), and the position of TRs has been elucidated by analysis of fully sequenced genomes. The data of these authors indicate firstly, that the density of TR array length (< 100 bp) varies widely in prokaryotic genomes (excess of repeats in Buchnera sp., Ureaplasma urealyticum). Secondly, it indicates that some species have a very strong excess of TR, with repeat unit lengths comprising a multiple of three (e.g. Mesorhizobium loti, Mycoplasma tuberculosis and Pseudomonas aeruginosa). A TR database is available at. This locates tandem repeats according to a number of criteria, among others, array length (length of gene fragment enclosed), repeat unit length and number of repeats.

A number of epidemiological molecular subtyping analysis studies have been published, with probably the most well known involving

Bacillus anthracis (Keim et al., 2000). Here, two different methods were used to enable typing of strains of B. anthracis, including those involved in bioterrorism. The methods were firstly, sequencing of the protective antigen gene on plasmid pXO1MLVApagA and secondly, Multi-Locus Variable-number tandem repeat Analysis (MLVA).

For MLVA analyses, 8 variable number TR markers (6 chromosomal markers and 2 plasmid markers), were combined in 4 multiplex PCR reactions. The PCR products were labelled with fluorescent primers using different coloured dyes, and the amplificates pooled for visualisation on an ABI Genescan. Thus, the location and length of amplificates were easy to detect and to evaluate by applying numerical analyses. The advantage of this system lies in the use of defined DNA regions across the entire genome.

The creation of a cumulative database has enabled the use of standardised amplification and separation methods. The disadvantage of this system lies in the need to have sequence information (preferably of more than a single strain of a species) available prior to the design of PCR primers.

With the combination of methods covering the isolation of DNA from any environmental sample (PCR amplification of 16S rRNA genes, cloning and sequence analysis), even as yet uncultured bacteria and microbial communities from natural samples have become accessible for identification and phylogenetic analysis. The presence of cells directly in the environment can be detected by in situ hybridisation techniques (FISH). These studies have demonstrated the enormous taxonomic, hence biotechnological, potential of hitherto unrecognised clostridia in nature.

Gas Chromatography of Metabolic End Products and Cellular Fatty Acids for Identifying Anaerobic Bacteria

Conventional methods of identifying anaerobic bacteria are based on morphology observation, culture and classical biochemical tests. In contrast to aerobes, these techniques are often insufficient to permit definitive identification, and additional investigations are required to complete the diagnosis. One of the most established methods consists of the detection and determination of organic compounds.

Two main groups of compounds may serve as support in the identification of anaerobes:

- Metabolites, produced by bacteria as a result of carbohydrate fermentation and/or breakdown of amino acids.

- Long-chain fatty acids, which are the major constituents of lipids (cell membranes), lipopolysaccharides (Gram-Negative bacteria) and lipoteichoic acid (Gram-Positive bacteria).

Gas chromatography (GC) is a well known technique allowing analysis of these metabolic or structural substances. Thus, GC offers 2 possibilities of additional means for the identification of anaerobes. In this chapter we examine the procedures and potential of each one of these approaches in anaerobic bacteriology. Anaerobic bacteria produce numerous metabolites, which mainly result from sugar fermentation and/or from the breakdown of amino acids by deamination, decarboxylation or Stickland reaction. These metabolites accumulate in the growth medium and are characteristic of bacterial genera and of some species.

Most of them have been described in the last 2 decades and consist of 2-oxo-acids, 2-hydroxy-acids, and phenolic compounds (Elsden and Hilton, 1976). Their detection by GC is the most common and most documented technique with regard to the identification of anaerobic bacteria. Nevertheless, some other analytical procedures including HPLC (High-Performance Liquid Chromatography), capillary electrophoresis and indirect UV detection (Arellano et al., 2000) have also been developed for the determination of volatile organic acids. From an analytical point of view, microbial metabolites are classified as volatile fatty acids (VFAs), also called short-chain fatty acids (SCFAs), and as nonvolatile fatty acids (NVFAs). In general, VFAs (acetic to heptanoic), which are the main metabolic products of interest in taxonomy, are directly analysed as free acids after extraction from the growth medium using either diethyl ether or methyl tert-butyl ether. The latter is a safer solvent, since it is considered less flammable.

Other methods have been developed, such as headspace chromatography, a complicated technique used for alcohols (C1–C4) and VFA analysis. Because peak areas are very dependent on analytical conditions (syringe and sample temperature), quantitative analysis is only possible with an automated GC system.

Direct analysis of the supernatants of acidified spent media, faecal or purulent material has also been proposed. A major drawback of this method is that non-volatile compounds tend to decompose in the injector and column. This problem can be solved by pre-treatment procedures or by use of a glass liner in the injector port. This liner is stopped with glass wool plugs and acts as a precolumn. In the last few years, an original extraction technique has been developed, which may be applied to the identification of anaerobic bacteria. Solid-phase microextraction (SPME) is a new solvent-free sample preparation technique, which is inexpensive and time efficient. It utilises a small, fusedsilica fibre coated with a thin layer of polymer to extract analytes from the sample matrix. Then the fibre is introduced into the gas chromatograph and the analytes are thermally desorbed into the injector and subsequently chromatographed. The analysis of fatty acids (C2–C10) with SPME can be carried out directly from liquid (broth culture) or from the headspace above the sample. Although this technique has not yet been applied to bacterial cultures, SPME seems to be a promising method for analysis of VFAs produced by anaerobes. Since this method allows for the making of derivatives, it is certainly also applicable to NVFA analysis. In conclusion, the use of SPME deserves to be developed in anaerobic microbiology.

NVFAs, such as oxo-acids, hydroxylated acids, dicarboxylic acids etc., have a high boiling point, and therefore cannot be directly detected by GC. They must undergo a derivatisation procedure prior to chromatographic analysis. The most commonly used method consists of esterification of analytes by direct incubation of media with methanol-hydrochloric or methanol-sulphuric acid followed by solvent extraction. However, the methylation reaction is incomplete and the recovery rate of methyl esters varies according to the acid. In addition, 2-oxo acids yield at least 2 products 133. A typical example is the methylation of pyruvic acid which produces methylpyruvate as expected, and a second component, identified as methyl-2,2-dimethoxypropanoate. In order to overcome these problems and to obtain reproducible accurate quantitative results, the method has been standardised and a known amount of internal standard (2-methylbenzoic acid) is added in advance to start the methylation reaction (Carlier, 1985).

Esterification of fatty acids can be carried out using different alcohols such as n-butanol, 2-butanol, propanol and methanol. A previous method consisted of converting VFAs and NVFAs to butyl esters with 14 % w/v boron trifluoride – butanol reagent. This method permitted separation of fatty acids (formic acid to heptanoic acid), lactic and succinic acids. Although this technique is no longer used in microbiology, it is always employed for the analysis of fatty acids in dairy products and can be very useful in accurate analysis of VFAs. Indeed, substituting the methyl group with a butyl group improves the flameionisation efficiency of fatty acid esters. Moreover, VFA butyl esters have a uniform flame-ionisation detection (FID) response, irrespective of the number of carbon atoms contained in the molecule

(Ulberth et al., 1999). Many other derivatisation procedures can be employed, but their description is beyond the scope of this chapter. (Review: Taguchi, 1990).

The analysis of metabolic end products can be utilised in 2 different ways: firstly, for a rapid presumptive diagnosis of anaerobic infection in the patient, or sometimes for identification by chemical marker of a particular species; secondly, as a tool in the identification of bacteria isolated from a clinical specimen. Proteolytic Clostridium species degrade amino acids (such as valine, leucine, isoleucine, phenylalanine, tyrosine and tryptophan), to iso acids and phenolic compounds. These latter compounds can be considered as microbial markers. Therefore, their detection by headspace analysis or direct injection of body fluids, acidified liquid or solid culture medium, indicates the presence of these proteolytic species in the sample. However, a microscopic examination of the sample is needed, since branched chain acids are also produced by Peptostreptococcus anaerobius and some Gram-Negative bacilli (Porphyromonas, Tissierella). In the same way, a presumptive identification of species can be encouraged by the detection of a specific compound after growth on supplemented broth culture. For example, C. dificile produces p-cresol on cycloserine-cefoxitin-fructose-egg yolk agar (CCFA) or pre-reduced anaerobically sterilised (PRAS) peptone yeast broth supplemented with tyrosine or p-134 hydroxyphenylacetic acid. C. malenominatum, C. tetani, C. tetanomorphum, C. cochlearium and C. limosum produce phenol from tyrosine (Elsden and Hilton, 1976).

Quantitative analysis of metabolic end products in conjunction with some simple tests such as gas, indole or catalase production, allows rapid identification of many common anaerobes. However, care must be taken to obtain reproducible results. Inoculum size, culture medium and time and condition of incubation must be standardised (Carlier, 1985). The more commonly isolated Clostridium may be subdivided into 2 chromatographic groups. The first is composed of species producing one or several straight chain volatile fatty acids, i.e. acetic, or acetic, propionic and butyric acids, whereas the second includes species generating both straight and branched-chain fatty acids.

If acetic acid is the only volatile acid produced, bacteria cannot be identified, since this chromatographic profile is common to many species. Consequently, biochemical tests are required for speciation except for C.135 clostridioforme, which often stains Gram-Negative and forms characteristic elongated cells. Among the species yielding butyric acid, C. perfringens produces propionic and 2-hydroxybutyric acids.

It is easy to recognise by its morphology and the presence of a double zone of haemolysis on a sheep-blood agar plate. Other species belonging to group 1 may be tentatively differentiated by some chromatographic characteristics. C. beijerinckii, unlike C. butyricum, does not produce acetic acid and only synthesises small amounts of lactic acid. C. innocuum exhibits a very specific profile, making its identification very easy without any other test. Two small peaks of acetate and caproate are always detected in addition to prominent peaks of butyric and lactic acids (Carlier, 2003). C. paraputrificum and C. tertium are very similar and must be identified on the basis of biochemical tests. C. tertium is nitrate-positive, whereas C. paraputrificum is not. C. symbiosum, a species morphologically similar to C. clostridioforme, is characterised by the production of butyric acid. Another related species, C. sphenoides, does not produce butyrate, but is indole positive (Holdeman et al., 1977). As seen above, 5 species possess chromatographic profiles similar to that of C. perfringens in their production of propionic and 2-hydroxybutyric acids.

However, they differ in terms of phenol production. Moreover, their cellular morphology is very different. In particular, these bacteria display terminal spores. Group 2 Distinctively, on the basis of aromatic and branched-chain acid production, the bacteria belonging to this group form 3 major subgroups. The first subgroup includes species producing isocaproate and one aromatic acid, namely 3-phenylpropionate, as differential metabolites. Note the higher production of 3-phenylpropionic acid associated with both C. sporogenes and proteolytic C. botulinum species. These species are phenotypically similar. In particular, they are lipase positive and proteolytic (milk-digested). Therefore, precise identification must be performed by a toxicity test. Neither related species, C. bifermentans nor C. sordellii, produces butyric acid, but they are both lecithinase positive and proteolytic. Interspecies differentiation can be carried out using the urease test, which is positive for C. sordellii. However, some urease negative strains have been described (Nakamura et al., 1976), and thus, these particular strains cannot be identified by GC pattern alone.

The second subgroup is formed by species producing mainly isocaproic acid and both aromatic acids, i.e. 2-phenylacetic and 3-phenylpropionic acids. C. mangenoti, is phenotypically very similar to the C. sporogenes/proteolytic C. botulinum group and also yields high levels of 3-phenylpropionic acid. However, it is lipase negative and non-toxigenic. This species is rarely present in clinical specimens.

C. dificile is easily identifiable by its characteristic metabolic end products. Indeed, almost all organic acids usually encountered in anaerobes are136 produced by this organism. The particular horse-stable odour, together with nonproteolytic activity complete identification.

The third subgroup comprises only 3 species: C. hastiforme, C. argentinense (formerly C. botulinum type G) and its non-toxigenic counterpart, C. subterminale. It is defined by the presence of isovaleric acid and the absence of 3-phenylpropionic acid. Also, much more importantly, production of 2-phenylacetic distinguishes this subgroup from the others. Production of propionic acid differentiates C. hastiforme from C. argentinense/C. subterminale.

Cellular fatty acid (CFA) composition of micro-organisms has been used extensively for bacterial identification and taxonomic purposes. Variability in number of carbon atoms, number and position of double-bonds, cis or trans isomers and substitution groups are stable parameters, provided that highly standardised analysis and culture conditions are observed.. The usual method consists of lysing bacterial cells by suspension in a saponification mixture, causing the release of their fatty acids, which are converted to methyl ester derivatives (FAME), and then analysed by gas chromatography. Esterification of fatty acids can be carried out by many other procedures involving acid or base-catalysed transesterification. In general, CFA analysis includes saturated, unsaturated, hydroxylated, straight and branched molecules with 9 to 20 carbon atoms. Identification is137 performed by comparison of sample retention times with those of commercially prepared FAME standard mixes. Tentative identification of unknown acids is achieved by measuring their equivalent chain length (ECL) against a homolog of the saturated straight-chain monocarboxylic acids.

The amount of each acid is determined via peak areas as a percentage of the total FAME peak areas. However, this technique is not practical for non-specialised laboratories. A major drawback involves the problem of reproducibility and of comparison of multivariate recordings. Besides, many other compounds other than FAMEs may be detected, making the chromatograms difficult to interpret. For example, many anaerobes contain plasmalogens, which are particular phospholipids possessing a vinyl ether bond at the sn-1 position, instead of the usual ester bond found in fatty acids. In this case, the methylation procedure leads to the formation of aldehydes, which are converted into dimethylacetals (DMAs). These derivatives are practically impossible to separate from most FAMEs. As a consequence, in addition to the need for a highly standardised procedure, the use of an automated apparatus coupled with a database is strongly recommended for effective interpretation of chromatograms and the accurate identification of micro-organisms.

Today, the only commercially available system for this purpose is the Microbial Identification System (MIS) produced by Microbial ID and distributed by Agilent Technologies. The MIS requires bacteria to be streaked on agar plates and incubated for 24 h before the derivatisation procedure and gas chromatography. The software takes into account both the presence or absence of FAMEs and DMAs and the ratio of the acids in the sample. The chromatographic profiles are compared with those contained in identification libraries by using a similarity index based upon calculation of the distance between the profile of the unknown strain and the mean profile of the closest library entry. A similarity value of 0.5 or greater is considered an acceptable identification, and a value below 0.5 is taken as an uncertain identification. However, the MIS is not very sensitive and 30– 40 mg of cells (wet weight) must be used in the analysis. This is a major drawback for fastidious or slow-growing organisms, requiring a long incubation period. Moreover, because CFA composition of bacteria varies with growth media and culture conditions, strict adherence to the analytical protocol recommended by the manufacturer must be followed to avoid misidentification (Ghanem et al., 1991).

Definition of ECL is as follows: For a specified column and carrier gas, in the range where there is a linear relationship between the molecular weight of the reference saturated straight-chain monocarboxylic methyl esters and logarithm of their retention times, interpolation between the retention times of the reference esters allows determination of an ECL for each compound. This value is independent of experimental conditions (Miwa et al., 1960).

In general, studies demonstrating the utility of CFA analysis by MIS for identification of Clostridium species are lacking. According to Allen et al. (1995), among 216 strains representing 18 species, MIS identified 86% of the strains correctly to species level without supplemental tests (C. perfringens, C. dificile, C. butyricum, C. sporogenes, C. septicum, C. glycolicum and C. symbiosum) and 6% of the strains were identified to species level with the aid of some complementary tests. 3% of the strains, including C. ramosum and

C. clostridioforme species, were misidentified to genus level. Another study showed that 100% of C. dificile strains (n=62) were correctly identified by the system, while misidentification was obtained for one strain of C. butyricum, one strain of C. sordellii and one strain of C. septicum (Anderson et al., 1995). From 686 analyses of 285 C. botulinum strains and related organisms, Ghanem et al. (1991) found 14 CFA groups for toxigenic strains and 6 other CFA groups for non-toxigenic strains. Depending on the toxin type or CFA group, the MIS system permitted a correct identification for 74 to 100% of the analyses, 89 to 100% for the toxigenic strains, and 60 to 100% for the non-toxigenic strains tested. The authors concluded that CFA analysis offers a reasonable alternative to mouse assays for laboratories that do not have facilities for these toxicity tests.

Demonstration of the presence of metabolites in spent media or headspace and analysis of CFA are 2 effective applications of GC for identification of bacteria. Detection of metabolic end products is the most used, simplest and least expensive method. Interpretation of chromatograms can be facilitated by comparison with data contained in numerous publications or by chromatography of reference strains. CFA analysis is an application requiring a highly standardised procedure. Furthermore, it is strongly dependent on the apparatus and software provided by the manufacturer. However, this technique allows the differentiation of more than 140 bacterial compounds (Ghanem et al., 1991), and is considered as an important feature in polyphasic taxonomy (Vandamme et al., 1996).

10

Fruit Processing Technology in Dairy

Enzymes are an integral component of modern fruit juice manufacturing and are highly suitable for optimising processes. Their main purposes are:

- increase extraction of juice from raw material
- increase processing efficiency (pressing, solid settling or removal)
- generate a final product that is clear and visually attractive.

Pomme Fruits

Depending on the chosen processing technology, different enzymes are used for mash treatment or extraction as well as for maceration. Juice depectinisation and starch degradation are important criteria for manufacturing concentrates.

The use of enzymes in fruit juice and cider processing: The juices from a wide variety of fruits such as apples, pears, mango and berries can be extracted to produce natural beverages. The processes used vary considerably depending on the type of fruit, its age and maturity. In general, extraction of fruit juice involves maceration followed by pressing or decanting, to separate the juice from the solids. Enzymes can play a key role in these processes improving yield, clarity and stability of the juice. Blends of macerating enzymes such as Macer8™ FJ and Pectinase 62L consisting of pectinases in combination with other carbohydrate degrading enzymes are useful in breaking open the fruit tissues to release more juice. Enzymatic maceration can result in extraction of >95% of the soluble solids from the fruit. Enzymes are also used to help clarify and stabilize the juice by degrading soluble pectins and starches that would otherwise cause haze.Biocatalysts provide a comprehensive range of enzymes to aid the production of fruit juice and cider.

The Fruit Cell Wall

The most important characteristic affecting the extraction of juice is the fruit cell wall, which provides the structure and rigidity of the pulp.

The composition of the cell wall can vary significantly for different types of fruit, but mainly consists of pectin, hemicellulose, cellulose, lignin and other components. The fruit cell wall has a highly complex structure. It is composed of bundles of crystalline cellulose microfibres embedded in an aqueous gel of hemicellulose and pectin. The xylan chains of the hemicellulose are cross-linked by phenolic acids (usually ferulic acid) and also linked to the arabinogalactan side chains of the pectin.

Fruit and vegetables are particularly rich in pectin which forms the major structural polysaccharide component of the fruit lamella and cell wall. It is composed of lengths of polygalacturonic acid (PGA) and rhamanogalactouronan 1 (RG1) with arabinan and galactan side chains. Pectins are often described as containing 'smooth' regions of linear PGA and 'hairy' regions of RG1 with multiple side chains. The linear chains of polygalacturonic acid (smooth regions) are largely esterified with methyl or acetyl groups. The extent of methylation affects the susceptibility of the pectin to pectinases and is characteristic of different types of fruit.

Apples, for example, usually have a very high pectin methylation.

Enzymes Used in Fruit Processing

Because fruit cell walls have such a complex structure of interwoven polymers, no single enzyme will breakdown cell walls. Complete liquefaction of a cell wall is likely to require a complex cocktail of carbohydrase enzymes. In fact, there is no single enzyme that will completely breakdown complex pectins. Commercial 'pectinases' are usually mixtures of enzymes such as polygalacturonase and pectin lyase. Biocatalysts range of macerating enzymes, for fruit juice processing, has been carefully tailored to contain the correct ratio of the enzymatic activities required for specific applications. The optimal activity of these enzymes is pH 4 - 5 (the natural pH of the fruit) and temperatures below 50°C. These are ideal conditions for fruit processing.

Polygalacturonase (PG)

Responsible for the random hydrolysis of alpha-1,4- glycosidic linkages between galacturonic acid residues. Depolymerise low esterified pectin (endo- and exo- enzymes) Pectin lyase (PL) - (EC 4.2.2.10), Cleaves the pectin, by an elimination reaction releasing oligosaccharides with non-reducing terminal alpha-1,4- linked galacturonic acid residues, without the necessity of pectin methyl esterase action Pectin methylesterase (PE) - (EC 3.2.1.11), Releases methanol from the pectyl methyl esters, a necessary stage before the polygalacturonase can act fully (the increase in the methanol content of such treated juice is generally less than the natural concentrations and poses no health risk).

Xylanase (Hemicellulase)

a mixture of hydrolytic enzymes including xylan endo-1,3-beta-xylosidase, (EC 3.2.1.32) and xylan 1,4-â-xylosidase, (EC 3.2.1.37), which degrade hemicellulose.

Arabonases (ARA)

Hydrolyse arabinans (EC 3.2.1.55).

Ferulic Acid Esterase (FAE)

Cuts ferulic acid and other phenolic linkages between the xylan chains opening the structure to further degradation by xylanases.

Cellulase

Breaks down cellulose (EC 3.2.1.4).

Amylases

Breaks down starch (EC 3.2.1.1).

Washing and Mechanical Maceration

The fruit is first washed to remove external surface dirt and topical chemical residues. It is then chopped and/or crushed to mechanically break open the tissues. This should be completed as quickly as possible to avoid browning reactions.

Enzymatic Maceration

Treatment of the mashed fruit with macerating enzymes such as Macer8™ FJ or Pectinase 62L further breaks down the fruit pulp resulting in increased yields of juice, reduced viscosity and improved run-off. The enzymes can be added in the crusher/pulper and may be diluted in water to assist homogenous distribution through the pulp.

The dose, time and temperature should be determined in trials depending on the degree of maceration required and the type of fruit. It is not usually necessary to adjust the pH of the pulp as Biocatalysts enzymes are designed to work at the natural pH of the fruit. Liquefaction may be achieved, although the enzymes are equally beneficial in yield improvement for juice extraction and viscosity reduction. Enzymatic maceration can also result in subtle but generally beneficial changes in the flavour and, in the case of cider-making, shorter fermentation times as valuable components are released.

Separation of Juice and Solids

The juice can be separated from the remaining solids in a press or in a decanter.

Secondary Extraction

To achieve the maximum utilization of the raw material, a secondary extraction using further enzymes may be carried out. The solids are reslurried with condensate, and after an appropriate residence time, extracted again. Using belt presses, yields of 92% of soluble solids are possible. A decanter instead of a belt press may possibly be an interesting alternative.

Clarification

One of the major problems encountered in the preparation of fruit juices and wine is cloudiness due primarily to the presence of pectins. One litre of juice with a dry matter content of 13% can contain 2-5g of pectin. The pectin can be associated with other plant polymers and the cell debris. The cloudiness that these cause is difficult to remove except by enzymic depectinization. After pressing, the juice is transferred to a stirred holding tank.

Pectinases such as Macer8™ FJ, Pectinase 62 L or Pectinase 444L can be added to the juice and incubated typically at 40°C-50°C. The dose should be determined in trials depending on the degree of depectinisation desired and the type of fruit used. It is not usually necessary to adjust the pH as Biocatalysts enzymes are designed to work at the natural pH of the fruit. The reaction is usually allowed to continue until a negative pectin (alcohol) test is obtained. The arabinase activity of preparations such as Pectinase 444L are particularly useful if the fruit contains high levels of arabans to prevent them forming a haze. As immature fruit can also contain high levels of starch it may be beneficial to also add a fungal amylase such as Depol 220L to prevent formation of haze in the pasteurisation stage. Enzymatic treatment has the additional benefits of reducing the solution viscosity. Insoluble plant material is easily removed by filtration, or settling and decantation, once the stabilising effect of the pectins on the colloidal haze has been removed.

Fining Agents (Bentonite)

The activity of pectinases are reduced by bentonite and it is recommended that enzyme activity is allowed to complete before fining agents are added.

Concentration/Pasteurization

The clarified fruit can then be concentrated, for example, by ultrafiltration and heat-treated (pasteurised) to kill any microorganisms that might be present. This heat treatment also helps improve the overall clarity of apple juice. Before being placed in the appropriate container (such as bottles), the juice may be further filtered and given an additional heat treatment to ensure safety.

Biocatalysts Enzymes for Juice and Cider Cellulase 13L/P;

- Suitable for use when complete cellulose hydrolysis is required e.g. fruit liquefaction
- Reduces viscosity
- Available as a powder or liquid
- Side activities of xylanase and glucanase.

Cellulase 13P is an enzyme preparation for the degradation of cellulose and other viscosity forming polysaccharides. Added to the crusher/pulper in combination with pectinases it can aid the complete liquefaction of fruits.

Macer8™ FJ;

- Gives improved performance in fruit juice extraction processes
- Successfully used with cranberries, strawberries and grapes
- Effective depectinisation due to high PG & PL
- Works at acidic pH (3.0 - 5.0)
- Low methanol production due to the low pectin esterase activity.

Macer8™ FJ is ideally suited for the production of high quality apple and pear juices. It contains a balanced mix of pectinases designed to provide controlled maceration and depectinisation of fruit, without generating undesirable levels of methanol. The controlled disintegration of the fruit results in improved juice yield and full flavour recovery.

Pectinase 444 L;

- Used to increase breakdown of cell wall components in processing
- Results in increased extraction of apple juice and decrease in waste
- Hydrolyses structural pectin to release shorter chain molecules which are more soluble and less fibrous
- High arabinase activity helps clarification and prevents haze
- Low PE ensures minimal methanol production
- Active down to pH 2.5 and from 10 to 60°C.

Pectinase 444L is a unique pectinase designed to provide economical depectinization of fruit juices. It can be added either to the pulp or juice to provide a controlled disintegration and rapid reduction in viscosity. It promotes improved juice yield, full flavour recovery and rapid clarification of the juice. It has an optimised PG to PL ratio and high maceration index. The high arabinase activity helps prevent arabinan haze formation and the low PE activity ensures only minimal levels of methanol are produced.

Depol 220 L;

- High strength fungal amylase
- Used to remove starch from apple juice
- Required when early season apples are used
- Eliminates the haze that can form on pasteurisation
- Increases stability of the product
- Active at low pH (3.0 - 6.0).

Depol 220L contains a high strength fungal alpha amylase used to prevent the formation of haze in juice made from early season apples. Although ripe fruit contains very little starch, immature fruit can contain up to 2% by weight of starch granules.

Following pasteurisation, this can produce a haze which is difficult to remove. Depol 220L can be added routinely at the clarification stage as a precaution against starch haze. It is active in the acidic conditions of apple juice. The additional sugars released sweeten the juice and can act as fermentable substrates in cider making.

Pectinase 62 L;

- Breakdown of structural pectins
- For use in white wine production
- Improves yield of grape must

- Enhances flavour by releasing desirable components from cell wall material
- High PG/PL ratio
- Arabanase activity.

Pectinase 62 L is a mixed pectinase preparation with a high PG to PL ratio and a moderate maceration index.

It also contains arabinase activity. It is used to improve the yield from pressed fruit pulps by breaking down pectins and can be used to clarify juice. It demonstrates excellent stability at low pH and is particularly suited for processing high-acidity fruit.

Glucose Oxidase G789L;

- Removal of oxygen from soft drinks
- Desugaring.

Glucose Oxidase G789L can be used to extend the shelf life of soft drinks by scavenging oxygen. The enzyme catalyses the oxidation of glucose with oxygen to form gluconate and hydrogen peroxide.

Exotic Fruit

Exotic fruit such as mangoes contain a high proportion of recalcitrant tissues which are not readily degraded by general fruit processing enzymes. These may require enzymes more typically associated with vegetable processing such as Depol™ 40L (D040L) and Depol™ 670L (D670L) for effective juice extraction.

Fruit and Vegetable Processing

In developing countries agriculture is the mainstay of the economy. As such, it should be no surprise that agricultural industries and related activities can account for a considerable proportion of their output. Of the various types of activities that can be termed as agriculturally based, fruit and vegetable processing are among the most important. Both established and planned fruit and vegetable processing projects aim at solving a very clearly identified development problem. This is that due to insufficient demand, weak infrastructure, poor transportation and perishable nature of the crops, the grower sustains substantial losses. During the post-harvest glut, the loss is considerable and often some of the produce has to be fed to animals or allowed to rot.

Even established fruit and vegetable canning factories or small/ medium scale processing centres suffer huge loss due to erratic supplies.

The grower may like to sell his produce in the open market directly to the consumer, or the produce may not be of high enough quality to process even though it might be good enough for the table. This means that processing capacities will be seriously underexploited.

The main objective of fruit and vegetable processing is to supply wholesome, safe, nutritious and acceptable food to consumers throughout the year. Fruit and vegetable processing projects also aim to replace imported products like squash, yams, tomato sauces, pickles, etc., besides earning foreign exchange by exporting finished or semi-processed products. The fruit and vegetable processing activities have been set up, or have to be established in developing countries for one or other of the following reasons:

- diversification of the economy, in order to reduce present dependence on one export commodity;
- government industrialisation policy;
- reduction of imports and meeting export demands;
- stimulate agricultural production by obtaining marketable products;
- generate both rural and urban employment;
- reduce fruit and vegetable losses;
- improve farmers' nutrition by allowing them to consume their own processed fruit and vegetables during the off-season;
- generate new sources of income for farmers/artisans;
- develop new value-added products.

Importance of Fruit and Vegetables in World Agriculture

Fruit and vegetables represent an important part of world agriculture production; some figures are seen in Table.

Table: *Fruit and Vegetable World Production, 1991*

Crop (Fruit)	***Production, 1000 T***	
	Total World	***Dev.ping all***
Appies	39404	14847
Apricots	2224	1147
Avocados	2036	1757
Bananas	47660	46753
Citrus fruits NES	1622	1231
Cantaloupes and other melons	12182	8733

Contd...

Crop (Fruit)	*Production, 1000 T*	
	Total World	*Dev.ping all*
Dates	3192	3146
Grapes	57188	14257
Grapefruit and pomelo	4655	2073
Lemons and limes	6786	4457
Mangoes	16127	16075
Oranges	55308	40325
Peaches and nectarines	8682	2684
Pears	9359	4431
Papayas	4265	4205
Plantains	26847	26847
Plums	5651	1806
Pineapples	10076	9183
Raisins	1041	470
Tangerines, mandarines, clementines	8951	4379
Watermelons	28943	19038
Currants	536009	
Raspberries	369087	
Strawberries	2469117	342009
Beans, green	3213	1702
Cabbages	36649	15569
Cauliflower	5258	2269
Carrots	13511	4545
Chilies + peppers, green	9145	6440
Cucumbers and gherkins	13619	7931
Eggplants	5797	4608
Garlic	3102	2446
Onions, dry	27977	17128
Peas, green	4856	1038
Pumpkins, squash, gourds	7933	6245

What Fruit and Vegetables can be Processed?

Practically any fruit and vegetable can be processed, but some important factors which determine whether it is worthwhile are:

a. the demand for a particular fruit or vegetable in the processed form;

b. the quality of the raw material, i.e. whether it can withstand processing;

c. regular supplies of the raw material.

For example, a particular variety of fruit which may be excellent to eat fresh is not necessarily good for processing. Processing requires frequent handling, high temperature and pressure.

Many of the ordinary table varieties of tomatoes, for instance, are not suitable for making paste or other processed products. A particular mango or pineapple may be very tasty eaten fresh, but when it goes to the processing centre it may fail to stand up to the processing requirements due to variations in its quality, size, maturity, variety and so on. Even when a variety can be processed, it is not suitable unless large and regular supplies are made available. An important processing centre or a factory cannot be planned just to rely on seasonal gluts; although it can take care of the gluts it will not run economically unless regular supplies are guaranteed.

To operate a fruit and vegetable processing centre efficiently it is of utmost importance to pre-organise growth, collection and transport of suitable raw material, either on the nucleus farm basis or using outgrowers.

Processing Planning

The secret of a well planned fruit and vegetable processing centre is that it must be designed to operate for as many months of the year as possible. This means the facilities, the buildings, the material handling and the equipment itself must be inter-linked and coordinated properly to allow as many products as possible to be handled at the same time, and yet the equipment must be versatile enough to be able to handle many products without major alterations.

A typical processing centre or factory should process four or five types of fruits harvested at different times of the year and two or three vegetables. This processing unit must also be capable of handling dried/dehydrated finished products, juices, pickles, tomato juice, ketchup and paste, jams, jellies and marmalades, semi-processed fruit products.

Advanced planning is necessary to process a large range of products in varied weather and temperature conditions, each requiring a special set of manufacturing and packaging formulae. The end result of the efforts should be a well-managed processing unit with lower initial investment.

A unit which is sensibly laid out and where one requirement co-relates to another, with a sound costing analysis, leads to an integrated operation. Instead of over-sophisticated machinery, a sensible simple processing unit may be required when planned production is not very large and is geared mainly to meet the demand of the domestic market.

Location

The basic objective is to choose the location which minimises the average production cost, including transport and handling. It is an advantage, all other things being equal, to locate a processing unit near the fresh raw material supply. It is a necessity for proper handling of the perishable raw materials, it allows the processing unit to allow the product to reach its best stage of maturation and lessens injury from handling and deterioration from changes during long transportation after harvesting.

An adequate supply of good water, availability of manpower, proximity to rail or road transport facilities and adequate markets are other important requirements.

Processing Systems

a. Small-Scale Processing. This is done by small-scale farmers for personal subsistence or for sale in nearby markets. In this system, processing requires little investment: however, it is time consuming and tedious. Until recently, small-scale processing satisfied the needs of rural and urban populations. However, with the rising rates of population and urbanisation growth and their more diversified food demands, there is need for more processed and diversified types of food.

b. Intermediate-Scale Processing. In this scale of processing, a group of small-scale processors pool their resources. This can also be done by individuals. Processing is based on the technology used by small-scale processors with differences in the type and capacity of equipment used. The raw materials are usually grown by the processors themselves or are purchased on contract from other farmers. These operations are usually located on the production site of in order to assure raw materials availability and reduce cost of transport. This system of processing can provide quantities of processed products to urban areas.

c. Large-Scale Processing. Processing in this system is highly mechanised and requires a substantial supply of raw materials for economical operation. This system requires a large capital investment and high technical and managerial skills. Because of the high demand for foods in recent years many large-scale factories were established in developing countries. Some succeeded, but the majority failed, especially in West Africa. Most of the failures were related to high labour inputs and relatively high cost, lack of managerial skills, high cost and supply instability of raw materials and changing governmental policies. Perhaps the most important reason for failure was lack of adequate quantity and regularity of raw material supply to factories. Despite the failure of these commercial operations, they should be able to succeed with better planning and management, along with the undertaking of more in-depth feasibility studies.

It can be concluded that all three types of processing systems have a place in developing countries to complement crop production to meet food demand. Historically, however, small and intermediate scale processing proved to be more successful than large-scale processing in developing countries.

Choice of Processing Technologies for Developing Countries

FAO maintains (in FAO, 1992c), that the basis for choosing a processing technology for developing countries ought to be to combine labour, material resources and capital so that not only the type and quantity of goods and services produced are taken into account, but also the distribution of their benefits and the prospects of overall growth. These should include:

a. increasing farmer/artisan income by the full utilisation of available indigenous raw material and local manufacturing of part or all processing equipment;
b. cutting production costs by better utilisation of local natural resources (solar energy) and reducing transport costs;
c. generating and distributing income by decentralising processing activities and involving different beneficiaries in processing activities (investors, newly employed, farmers and small-scale industry);
d. maximising national output by reducing capital expenditure and royalty payments, more effectively developing balance-of-payments deficits through minimising imports (equipment, packing material, additives), and maximising export-oriented production;

e. maximising availability of consumer goods by maximisation of high-quality, standard processed produce for internal and export markets, reducing post-harvest losses, giving added value to indigenous crops and increasing the volume and quality of agricultural output.

Knowledge and control of the means of production, local manufacturing of processing equipment and development of appropriate/new technologies and more suitable raw material for processing must all be better researched. Decentralisation of activities must be maintained and coordinated.

The introduction of more sophisticated processing equipment and packaging material must be subordinated to internal and export marketing references.

Choosing a technology solely to maximise profits can actually work against true development. Choice should also be based on a solid, long-term market opportunity to ensure viability.

The internal market should be given greater consideration, safeguarded and supported. Training courses, at all levels, in processing and preservation of indigenous crops, must be expanded.

General Properties of Fruit and Vegetables; Chemical Composition and Nutritional Aspects; Structural Features

Fruit and vegetables have many similarities with respect to their compositions, methods of cultivation and harvesting, storage properties and processing. In fact, many vegetables may be considered fruit in the true botanical sense. Botanically, fruits are those portions of the plant which house seeds.

Therefore such items as tomatoes, cucumbers, eggplant, peppers, and others would be classified as fruits on this basis. However, the important distinction between fruit and vegetables has come to be made on an usage basis. Those plant items that are generally eaten with the main course of a meal are considered to be vegetables.

Those that are commonly eaten as dessert are considered fruits. That is the distinction made by the food processor, certain marketing laws and the consuming public, and this distinction will be followed in this document.

Vegetables are derived from various parts of plants and it is sometimes useful to associate different vegetables with the parts of the plant they represent since this provides clues to some of the characteristics we may expect in these items. A classification of vegetables based on morphological features is seen in Table.

Table: *Classification of Vegetables*

Category	***Examples***
Earth vegetables roots	sweet potatoes, carrots
modified stems tubers	potatoes
modified buds bulbs	onions, garlic
Herbage vegetables	
leaves	cabbage, spinach, lettuce
petioles (leaf stalk)	celery, rhubarb
flower buds	cauliflower, artichokes
sprouts, shoots (young stems)	asparagus, bamboo shoots
Fruit vegetables	
legumes	peas, green beans
cereals	sweet corn
vine fruits	squash, cucumber
berry fruits	tomato, egg plant
tree fruits	avocado, breadfruit

Source: Feinberg (1973)

Fruit as a dessert item, is the mature ovaries of plants with their seeds. The edible portion of most fruit is the fleshy part of the pericarp or vessel surrounding the seeds. Fruit in general is acidic and sugary. They commonly are grouped into several major divisions, depending principally upon botanical structure, chemical composition and climatic requirements.

Berries are fruit which are generally small and quite fragile. Grapes are also physically fragile and grow in clusters. Melons, on the other hand, are large and have a tough outer rind. Drupes (stone fruit) contain single pits and include such items as apricots, cherries, peaches and plums. Pomes contain many pits, and are represented by apples, quince and pears.

Citrus fruit like oranges, grapefruit and lemons are high in citric acid. Tropical and subtropical fruits include bananas, dates, figs, pineapples, mangoes, and others which require warm climates, but exclude the separate group of citrus fruits.

The compositions of representative vegetables and fruits in comparison with a few of the cereal grains are seen in Table.

Table: *Typical percentage composition of foods from plant origin*
Percentage Composition- Edible Portion

Food	***Carbo-hydrate***	***Protein***	***Fat***	***Ash***	***Water***
Cereals					
wheat flour, white	73.9	10.5	1.9	1.7	12
rice, milled, white	78.9	6.7	0.7	0.7	13
maize, whole grain	72.9	9.5	4.3	1.3	12
Earth vegetables					
potatoes, white	18.9	2.0	0.1	1.0	78
sweet potatoes	27.3	1.3	0.4	1.0	70
Vegetables					
carrots	9.1	1.1	0.2	1.0	88.6
radishes	4.2	1.1	0.1	0.9	93.7
asparagus	4.1	2.1	0.2	0.7	92.9
beans, snap, green	7.6	2.4	0.2	0.7	89.1
peas, fresh	17.0	6.7	0.4	0.9	75.0
lettuce	2.8	1.3	0.2	0.9	94.8
Fruit					
banana	24.0	1.3	0.4	0.8	73.5
orange	11.3	0.9	0.2	0.5	87.1
apple	15.0	0.3	0.4	0.3	84.0
strawberries	8.3	0.8	0.5	0.5	89.9

Source: Anon. (196O)

Compositions of vegetables and fruit not only vary for a given kind in according to botanical variety, cultivation practices, and weather, but change with the degree of maturity prior to harvest, and the condition of ripeness, which is progressive after harvest and is further influenced by storage conditions. Nevertheless, some generalisations can be made. Most fresh vegetables and fruit are high in water content, low in protein, and low in fat. In these cases water contents will generally be greater than 70% and frequently greater than 85%.

Commonly protein content will not be greater than 3.5% or fat content greater than 0.5 %. Exceptions exist in the case of dates and raisins which are substantially lower in moisture but cannot be considered fresh in the same sense as other fruit. Legumes such as peas and certain beans are higher in protein; a few vegetables such as sweet corn which are slightly higher in fat and avocados which are substantially higher in fat.

Vegetables and fruit are important sources of both digestible and indigestible carbohydrates. The digestible carbohydrates are present largely in the form of sugars and starches while indigestible cellulose provides roughage which is important to normal digestion.

Fruit and vegetables are also important sources of minerals and certain vitamins, especially vitamins A and C. The precursors of vitamin A, including beta-carotene and certain other carotenoids, are to be found particularly in the yellow-orange fruit and vegetables and in the green leafy vegetables.

Citrus fruit are excellent sources of vitamin C, as are green leafy vegetables and tomatoes. Potatoes also provide an important source of vitamin C for the diets of many countries. This is not so much due to the level of vitamin C in potatoes which is not especially high but rather to the large quantities of potatoes consumed.

Chemical Composition

Water

Vegetal cells contain important quantities of water. Water plays a vital role in the evolution and reproduction cycle and in physiological processes. It has effects on the storage period length and on the consumption of tissue reserve substances.

In vegetal cells, water is present in following forms:

- bound water or dilution water which is present in the cell and forms true solutions with mineral or organic substances;
- colloidal bound water which is present in the membrane, cytoplasm and nucleus and acts as a swelling agent for these colloidal structure substances; it is very difficult to remove during drying/dehydration processes;
- constitution water, directly bound on the chemical component molecules and which is also removed with difficulty.

Vegetables contain generally 90-96% water while for fruit normal water content is between 80 and 90%.

Mineral Substances

Mineral substances are present as salts of organic or inorganic acids or as complex organic combinations (chlorophyll, lecithin, etc.); they are in many cases dissolved in cellular juice.

Vegetables are more rich in mineral substances as compared with fruits. The mineral substance content is normally between 0.60 and 1.80% and more than 60 elements are present; the major elements are: K, Na, Ca, Mg, Fe, Mn, Al, P. Cl, S.

Among the vegetables which are especially rich in mineral substances are: spinach, carrots, cabbage and tomatoes. Mineral rich fruit includes: strawberries, cherries, peaches and raspberries. Important quantities of potassium (K) and absence of sodium chloride (NaCl) give a high dietetic value to fruit and to their processed products. Phosphorus is supplied mainly by vegetables.

Vegetables usually contain more calcium than fruit; green beans, cabbage, onions and beans contain more than 0.1% calcium. The calcium/phosphorus or Ca/P ratio is essential for calcium fixation in the human body; this value is considered normal at 0.7 for adults and at 1.0 for children. Some fruit are important for their Ca/P ratio above 1.0: pears, lemons, oranges and some temperate climate mountain fruits and wild berries.

Even if its content in the human body is very low, iron (Fe) has an important role as a constituent of haemoglobin. Main iron sources are apples and spinach. Salts from fruit have a basic reaction; for this reason fruit consumption facilitates the neutralisation of noxious uric acid reactions and contributes to the acid-basic equilibrium in the blood.

Carbohydrates

Carbohydrates are the main component of fruit and vegetables and represent more than 90% of their dry matter. From an energy point of view carbohydrates represent the most valuable of the food components; daily adult intake should contain about 500 g carbohydrates. Carbohydrates play a major role in biological systems and in foods. They are produced by the process of photosynthesis in green plants. They may serve as structural components as in the case of cellulose; they may be stored as energy reserves as in the case of starch in plants; they may function as essential components of nucleic acids as in the case of ribose; and as components of vitamins such as ribose and riboflavin.

Carbohydrates can be oxidised to furnish energy, and glucose in the blood is a ready source of energy for the human body. Fermentation of carbohydrates by yeast and other microorganisms can yield carbon dioxide, alcohol, organic acids and other compounds.

Some properties of sugars. Sugars such as glucose, fructose, maltose and sucrose all share the following characteristics in varying degrees, related to fruit and vegetable technology:

- they supply energy for nutrition;
- they are readily fermented by micro-organisms;
- in high concentrations they prevent the growth of micro-organisms, so they may be used as a preservative;
- on heating they darken in colour or caramelise;
- some of them combine with proteins to give dark colours known as the browning reaction.

Some properties of starches:

- They provide a reserve energy source in plants and supply energy in nutrition;
- they occur in seeds and tubers as characteristic starch granules.

Some properties of celluloses and hemicelluloses:

- They are abundant in the plant kingdom and act primarily as supporting structures in the plant tissues;
- they are insoluble in cold and hot water;
- they are not digested by man and so do not yield energy for nutrition;
- the fibre in food which produces necessary roughage is largely cellulose.

Some properties of pectins and carbohydrate gums.

- Pectins are common in fruits and vegetables and are gum-like (they are found in and between cell walls) and help hold the plant cells together;
- pectins in colloidal solution contribute to viscosity of the tomato paste;
- pectins in solution form gels when sugar and acid are added; this is the basis of jelly manufacture.

Fats

Generally fruit and vegetables contain very low level of fats, below 0.5%.

However, significant quantities are found in nuts (55%), apricot kernel (40%), grapes seeds (16%), apple seeds (20%) and tomato seeds (18%).

Organic Acids

Fruit contains natural acids, such as citric acid in oranges and lemons, malic acid of apples, and tartaric acid of grapes. These acids give the fruits tartness and slow down bacterial spoilage. We deliberately ferment some foods with desirable bacteria to produce acids and this give the food flavour and keeping quality. Examples are fermentation of cabbage to produce lactic acid and yield sauerkraut and fermentation of apple juice to produce first alcohol and then acetic acid to obtain vinegar. Organic acids influence the colour of foods since many plant pigments are natural pH indicators.

With respect to bacterial spoilage, a most important contribution of organic acids is in lowering a food's pH. Under anaerobic conditions and slightly above a pH of 4.6, Clostridium botulinum can grow and produce lethal toxins. This hazard is absent from foods high in organic acids resulting in a pH of 4.6 and less. Acidity and sugars are two main elements which determine the taste of fruit. The sugar/acid ratio is very often used in order to give a technological characterisation of fruits and of some vegetables.

Nitrogen-containing Substances

These substances are found in plants as different combinations: proteins, amino acids, amides, amines, nitrates, etc. Vegetables contain between 1.0 and 5.5 % while in fruit nitrogen-containing substances are less than 1% in most cases. Among nitrogen containing substances the most important are proteins; they have a colloidal structure and, by heating, their water solution above 50°C an one-way reaction makes them insoluble. This behaviour has to be taken into account in heat processing of fruits and vegetables.

From a biological point of view vegetal proteins are less valuable then animal ones because in their composition all essential amino-acids are not present.

Vitamins

Vitamins are defined as organic materials which must be supplied to the human body in small amounts apart from the essential amino-acids or fatty acids. Vitamins function as enzyme systems which facilitate the metabolism of proteins, carbohydrates and fats but there is growing evidence that their roles in maintaining health may extend yet further.

The vitamins are conveniently divided into two major groups, those that are fat-soluble and those that are water-soluble. Fat-soluble vitamins are A, D, E and K. Their absorption by the body depends upon the normal absorption of fat from the diet. Water-soluble vitamins include vitamin C and several members of the vitamin B complex.

Vitamin A or Retinol

This vitamin is found as such only in animal materials - meat, milk, eggs and the like. Plants contain no vitamin A but contain its precursor, beta-carotene. Man needs either vitamin A or beta-carotene which he can easily convert to vitamin A. Beta-carotene is found in the orange and yellow vegetables as well as the green leafy vegetables, mainly carrots, squash, sweet potatoes, spinach and kale.

A deficiency of vitamin A leads to night blindness, failure of normal bone and tooth development in the young and diseases of epithelial cells and membrane of the nose, throat and eyes which decrease the body's resistance to infection.

Vitamin C

Vitamin C is the anti-scurvy vitamin. Lack of it causes fragile capillary walls, easy bleeding of the gums, loosening of teeth and bone joint diseases. It is necessary for the normal formation of the protein collagen, which is an important constituent of skin and connective tissue. Like vitamin E, vitamin C favours the absorption of iron.

Vitamin C, also known as ascorbic acid, is easily destroyed by oxidation especially at high temperatures and is the vitamin most easily lost during processing, storage and cooking. Excellent sources of vitamin C are citrus fruits, tomatoes, cabbage and green peppers. Potatoes also are a fair source (although the content of vitamin C is relatively low) because we consume large quantities of potatoes.

Enzymes

Enzymes are biological catalysts that promote most of the biochemical reactions which occur in vegetable cells.

Some properties of enzymes important in fruit and vegetable technology are the following:

- in living fruit and vegetables enzymes control the reactions associated with ripening;
- after harvest, unless destroyed by heat, chemicals or some other means, enzymes continue the ripening process, in many cases to the point of spoilage - such as soft melons or overripe bananas;

- because enzymes enter into a vast number of biochemical reactions in fruits and vegetable, they may be responsible for changes in flavour, colour, texture and nutritional properties;
- the heating processes in fruit and vegetables manufacturing/ processing are designed not only to destroy micro-organisms but also to deactivate enzymes and so improve the fruit and vegetables' storage stability.

Enzymes have an optimal temperature - around +50°C where their activity is at maximum. Heating beyond this optimal temperature deactivates the enzyme. Activity of each enzyme is also characterised by an optimal pH. In fruit and vegetable storage and processing the most important roles are played by the enzymes classes of hydrolases (lipase, invertase, tannase, chlorophylase, amylase, cellulase) and oxidoreductases (peroxidase, tyrosinase, catalase, ascorbinase, polyphenoloxidase).

Turgidity and Texture

The range of textures that are encountered in fresh and cooked vegetables and fruit is indeed great, and to a large extent can be explained in terms of changes in specific cellular components. Since plants tissues generally contain more than two-thirds water, the relationships between these components and water further determine textural differences.

Cell Turgidity. - Quite apart from other contributing factors, the state of turgidity, determined by osmotic forces, plays a paramount role in the texture of fruit and vegetables. The cell walls of plant tissues have varying degrees of elasticity and are largely permeable to water and ions as well as to small molecules.

The membranes of the living protoplast are semi-permeable, that is they allow passage of water but are selective with respect to transfer of dissolved and suspended materials. The cell vacuoles contain most of the water in plant cells and sugars, acids, salts, amino acids, some water-soluble pigments and vitamins, and other low molecular weight constituents are dissolved in this water.

In the living plant, water taken up by the roots passes through the cell walls and membranes into the cytoplasm of the protoplasts and into the vacuoles to establish a state of osmotic equilibrium within the cells. The osmotic pressure within the cell vacuoles and within the protoplasts pushes the protoplasts against the cell walls and causes them to stretch slightly in accordance with their elastic properties. This is the situation in the growing plant and the harvested live fruit or vegetable which is responsible for desired plumpness, succulence, and much of the crispness.

When plant tissues are damaged or killed by storage, freezing, cooking, or other causes, an important major change that results is denaturation of the proteins of cell membranes resulting in the loss of perm-selectivity. Without perm-selectivity the state of osmotic pressure in cell vacuoles and protoplasts cannot exist, and water and dissolved substances are free to diffuse out of the cells and leave the remaining tissue in a soft and wilted condition.

Other Factors Affecting Texture. The existence of a high degree of turgidity in live fruit and vegetables or whether a relative state of flabbiness develops from loss of osmotic pressure as well as final texture depends on several cell constituents. Cellulose, Hemicellulose, and Lignin. Cell walls in young plants are very thin and are composed largely of cellulose. As the plant ages cell walls tend to thicken and become higher in hemicellulose and in lignin. These materials are fibrous and tough and are not significantly softened by cooking.

Pectic Substances. The complex polymers of sugar acid derivatives include pectin and closely related substances. The cement-like substance found especially in the middle lamella which helps hold plant cells to one another is a water-insoluble pectic substance.

On mild hydrolysis it yields water-soluble pectin which can form gels or viscous colloidal suspensions with sugar and acid. Certain water-soluble pectic substances also react with metal ions, particularly calcium, to form water-insoluble salts such as calcium pectates. The various pectic substances may influence texture of vegetables and fruits in several ways.

When vegetables or fruit are cooked, some of the water-insoluble pectic substance is hydrolysed into water-soluble pectin. This results in a degree of cell separation in the tissues and contributes to tenderness. Since many fruits and vegetables are somewhat acidic and contain sugars the soluble pectin also tends to form colloidal suspensions which will thicken the juice or pulp of these products.

Fruit and vegetables also contain a natural enzyme which can further hydrolyse pectin to the point where the pectin loses much of its gel forming property. This enzyme is known as pectin methyl esterase. Materials such as tomato juice or tomato paste will contain both pectin and pectin methyl esterase. If freshly prepared tomato juice or paste is allowed to stand the original viscosity gradually decreases due to the action of pectin methyl esterase on pectin gel. This can be prevented if the tomato products are quickly heated to a temperature of about 82°C (180 F°) to deactivate the pectin methyl esterase liberated from broken cells before it has a chance to hydrolyse the pectin. Such a treatment is commonly practiced in the manufacture of tomato juice products. This is known as the "hot-break process" and yields products of high viscosity.

In contrast, where low viscosity products are desired no heat is used and enzyme activity is allowed to proceed. This is "cold-break" process. After sufficient decrease in viscosity is achieved the product can be heat treated, as in canning, to preserve it for long term storage.

It is often also desirable to firm the texture of fruit and vegetables, especially when products are normally softened by processing. In this case advantage is taken of the reaction between soluble pectic substances and calcium ions which form calcium pectates. These calcium pectates are water insoluble and when they are produced within the tissues of fruit and vegetables they increase structural rigidity. Thus, it is common commercial practice to add low levels of calcium salts to tomatoes, apples, and other vegetables and fruits prior to canning or freezing.

Sources of Colour and Colour Changes

In addition to a great range of textures, much of the interest that fruits and vegetables add to our diets is due to their delightful and variable colours. The pigments and colour precursors of fruit and vegetables occur for the most part in the cellular plastic inclusions such as the chloroplasts and other chromoplasts, and to a lesser extent dissolved in fat droplets or water within the cell protoplast and vacuoles. These pigments are classified into four major groups which include the chlorophylls, carotenoids, anthocyanins, and anthoanthins. Pigments belonging to the latter two groups also are referred to as flavonoids, and include the tannins.

The Chlorophylls. The chlorophylls are contained mainly within the chloroplasts and have a primary role in the photosynthetic production of carbohydrates from carbon dioxide and water. The bright green colour of leaves and other parts of plants is largely due to the oilsoluble chlorophylls, which in nature are bound to protein molecules in highly organised complexes. When the plant cells are killed by ageing, processing, or cooking, the protein of these complexes is denatured and the chlorophyll may be released. Such chlorophyll is highly unstable and rapidly changes in colour to olive green and brown. This colour change is believed to be due to the conversion of chlorophyll to the compound pheophytin.

Conversion to pheophytin is favoured by acid pH but does not occur readily under alkaline conditions. For this reason peas, beans, spinach, and other green vegetables which tend to lose their bright green colours on heating can be largely protected against such colour changes by the addition of sodium bicarbonate or other alkali to the cooking or canning water. However, this practice is not looked upon favourably nor used commercially because alkaline pH also has a softening effect on cellulose and vegetable texture and also destroys vitamin C and thiamin at cooking temperatures.

The Carotenoids. Pigments belonging to this group are fat-soluble and range in colour from yellow through orange to red. They often occur along with the chlorophylls in the chloroplasts, but also are present in other chromoplasts and may occur free in fat droplets. Important carotenoids include the orange carotenes of carrot, maize, apricot, peach, citrus fruits, and squash; the red lycopene of tomato, watermelon, and apricot; the yellow-orange xanthophyll of maize, peach, paprika and squash; and the yellow-orange crocetin of the spice saffron. These and other carotenoids seldom occur singly within plant cells.

A major importance of some of the carotenoids is their relationship to vitamin A. A molecule of orange beta-carotene is converted into two molecules of colourless vitamin A in the animal body. Other carotenoids like alpha-carotene, gamma-carotene, and cryptoxanthin also are precursors of vitamin A, but because of minor differences in chemical structure one molecule of each of these yields only one molecule of vitamin A. In food processing the carotenoids are fairly resistant to heat, changes in pH, and water leaching since they are fat-soluble. However, they are very sensitive to oxidation, which results in both colour loss and destruction of vitamin A activity.

The Flavonoids. Pigments and colour precursors belonging to this class are water-soluble and commonly are present in the juices of fruit and vegetables. The flavonoids include the purple, blue, and red anthocyanins of grapes, berries, plump, eggplant, and cherry; the yellow anthoxanthins of light coloured fruit and vegetables such as apple, onion, potato, and cauliflower, and the colourless catechins and leucoanthocyanins which are food tannins and are found in apples, grapes, tea, and other plant tissues. These colourless tannin compounds are easily converted to brown pigments upon reaction with metal ions. Properties of the anthocyanins include a shifting of colours with pH.

Thus many of the anthocyanins which are violet or blue in alkaline media become red upon addition of acid. Cooking of beets with vinegar tends to shift the colour from a purplish red to a brighter red, while alkaline water can influence the colour of red fruits and vegetables toward violet and gray-blue.

The anthocyanins also tend toward the violet and blue hues upon reaction with metal ions, which is one reason for lacquering the inside of metal cans when the true colour of anthocyanin-containing fruits and vegetables is to be preserved. The water-soluble property of anthocyanins also results in easy leaching of these pigments from cut fruit and vegetables during processing and cooking. The yellow anthoxanthins also are pH sensitive tending toward a deeper yellow in alkaline media. Thus potatoes or apples become somewhat yellow when cooked in water with a pH of 8 or higher, which is common in many areas. Acidification of the water to pH 6 or lower favours a whiter colour.

The colourless tannin compounds upon reaction with metal ions form a range of dark coloured complexes which may be red, brown, green, grey, or black. The various shades of these coloured complexes depend upon the particular tannin, the specific metal ion, pH, concentration of the complex, and other factors not yet fully understood.

Water-soluble tannins appear in the juices squeezed from grapes, apples, and other fruits as well as the brews from extraction of tea and coffee. The colour and clarity of tea are influenced by the hardness and pH of the brewing water. Alkaline waters that contain calcium and magnesium favour the formation of dark brown tannin complexes which precipitate when the tea is cooled. If acid in the form of lemon juice is added to such tea its colour lightens and the precipitate tends to dissolve. Iron from equipment or from pitted tin cans has caused a number of unexpected colours to develop in products containing tannins, such as coffee, cocoa and foods flavoured with these. The tannins are also important because they have an astringency which influences flavour and contributes body to such beverages as tea, wine, apple cider, etc.

Activities of Living Systems

Fruit and vegetables are in a live state after harvest. Continued respiration gives off carbon dioxide, moisture, and heat which influence storage, packaging, and refrigeration requirements. Continued transpiration adds to moisture evolved and further influences packaging requirements.

Further activities of fruit and vegetables, before and after harvest, include changes in carbohydrates, pectins, organic acids, and the effects these have on various quality attributes of the products. As for changes in carbohydrates, few generalizations can be given with respect to starches and sugars. In some plant products sugars quickly decrease and starch increases in amount soon after harvest. This is the case for ripe sweet corn which can suffer flavour and texture quality losses in a very few hours after harvest.

Unripe fruit, in contrast, is frequently high in starch and low in sugars. Continued ripening after harvest generally results in a decrease in starch and a increase in sugars as in the case of apples and pears. However, this does not necessarily mean that the starch is the source of the newly formed sugars.

Further, the courses of change in starch and sugars are markedly influenced by postharvest storage temperatures. Thus potatoes stored below about 10 C° (50 F°) continue to build up high levels of sugars, while the same potatoes stored above 10 C° do not. This property is used to help the dehydration process in potato storage. Here potatoes should have a low reducing sugar content so as to minimise Maillard browning reactions during drying and subsequent storage of the dried product. In this case potatoes are stored above 10°C prior to being further processed. After harvest the pectin changes in fruit and vegetables are more predictable. Generally there is decrease in water-insoluble pectic substance and a corresponding increase in watersoluble pectin. This contributes to the gradual softening of fruits and vegetables during storage and ripening. Further breakdown of water-soluble pectin by pectin methyl esterase also occurs.

The organic acids of fruit generally decreases during storage and ripening. This occurs in apples and pears and is especially important in the case of oranges. Oranges have a long ripening period on the tree and time of picking is largely determined by degree of acidity and sugar content which have major effects upon juice quality. It is important to note that the reduction of acid content on ripening influences more than just the tartness of fruit. Since many of the plant pigments are sensitive to acid, fruit colour would be expected to change. Additionally, the viscosity of pectin gel is affected by acid and sugar contents, both of which change with ripening.

Stability of Nutrients

One of the principal responsibilities of the food scientist and food technologist is to preserve food nutrients through all phases of food acquisition, processing, storage, and preparation. The key is in the specific sensitivities of the various nutrients, the principles of which are illustrated in Table.

Table: *Specific sensitivity and stability of nutrients**

Nutrient	*Neutral pH 7*	*Acid < pH 7*	*Alkaline > pH 7*	*Air or Oxygen*	*Light*	*Heat*	*Cooking Losses, Range*
Vitamins							
Vitamin A	S	U	S	U	U	U	0-40
Ascorbic acid(C)	U	S	U	U	U	U	0-100
Biotin	S	S	S	S	S	U	0-60
Carotenes	S	U	S	U	U	U	0-30
(pro A)							0-5
Choline	S	S	S	U	S	S	0-10
Cobalamin (B12)	S	S	S	U	U	S	0-40
Vitamin D	S		U	U	U	U	0-10
Essential fatty acids	S	S	U	U	U	S	
Folic acid	U	U	S	U	U	U	0-100
Inositol	S	S	S	S	S	U	0-95
Vitamin K	S	U	U	S	U	S	0-5
Niacin (PP)	S	S	S	S	S	S	0-75
Pantbothenic acid	S	U	U	S	S	U	0-50
p-Amino Benzoic acid	S	S	S	U	S	S	0-5
Vitamin B6	S	S	S	S	U	U	0-40
Riboflavin (b) 4	S	S	U	S	U	U	0-75
Tbiamin (b) 4	U	S	U	U	S	U	0-80
Tocopherols	S	S	S	U	U	U	0-55

*Source: Harris and Karmas, 1975

(U = Unstable; S = Stable)

This shows the stability of vitamins, essential amino acids, and minerals to acid, air, light, and heat, and gives an indication of possible cooking losses. Vitamin A is highly sensitive to acid, air, light and heat; vitamin C to alkalinity, air, light and heat; vitamin D to alkalinity, air, light and heat; thiamin to alkalinity, air, and heat in alkaline solutions; etc. Cooking losses of some essential nutrients may be in excess of 75%. In modern food processing operations, however, losses are seldom in excess of 25%.

The ultimate nutritive value of a food results from the sum total of losses incurred throughout its history - from farmer to consumer. Nutrient value begins with genetics of the plant and animal. The farmland fertilization program affects tissue composition of plants, and animals consuming these plants. The weather and degree of maturity at harvest affect tissue composition.

Storage conditions before processing affect vitamins and other nutrients. Washing, trimming, and heat treatments affect nutrient content. Canning, evaporating, drying, and freezing alter nutritional values, and the choices of times and temperatures in these operations frequently must be balanced between good bacterial destruction and minimum nutrient destruction.

Packaging and subsequent storage affect nutrients. One of the most important factors is the final preparation of the food in the home and the restaurant - the steam table can destroy much of what has been preserved through all prior manipulations.

Structural Features

The structural unit of the edible portion of most fruits and vegetables is the parenchyma cell. While parenchyma cells of different fruit and vegetables differ somewhat in gross size and appearance, all have essentially the same fundamental structure. Parenchyma cells of plants differ from animal cells in that the actively metabolising protoplast portion of plant cells represents only a small fraction, of the order of five per cent, of the total cell volume. This protoplast is film-like and is pressed against the cell wall by the large water-filled central vacuole.

The protoplast has inner and outer semi-permeable membrane layers; the cytoplasm and its nucleus are held between them. The cytoplasm contains various inclusions, among them starch granules and plastics such as the chloroplasts and other pigment-containing chromoplasts.

The cell wall, cellulose in nature, contributes rigidity to the parenchyma cell and limits the outer protoplasmic membrane. It is also the structure against which other parenchyma cells are cemented to form extensive three-dimensional tissue masses.

The layer between cell walls of adjacent parenchyma cells is referred to as the middle lamella, and is composed largely of pectic and polysaccharide cement-like materials. Air spaces also exist, especially at the angles formed where several cells come together.

The relationships between these structures and their chemical compositions are further outlined below.

The parenchyma cells will vary in size among plants but are quite large when compared to bacterial or yeast cells. The larger parenchyma cells may have volumes many thousand times greater than a typical bacterial cell.

There are additional types of cells other than parenchyma cells that make up the familiar structures of fruit and vegetables.

These include various types of conducting cells which are tube-like and distribute water and salts throughout the plant. Such cells produce fibrous structures toughened by the presence of cellulose and the woodlike substance lignin. Cellulose, lignin, and pectic substances also occur in specialised supporting cells which increase in importance as plants become older. An important structural feature of all plants, including fruit and vegetables is protective tissue. This can take many forms but usually is made up of specialised parenchyma cells that are pressed compactly together to form a skin, peel or rind.

Surface cells of these protective structures on leaves, stems or fruit secrete waxy cutin and form a water impermeable cuticle. These surface tissues, especially on leaves and young stems will also contain numerous valve-like cellular structures, the stomata, through which moisture and gases can pass.

Deterioration Factors and their Control

A summary of overall deterioration reactions in fruits and vegetables is presented below.

Enzymic Changes

Enzymes which are endogenous to plant tissues can have undesirable or desirable consequences. Examples involving endogenous enzymes include

a) the post-harvest senescence and spoilage of fruit and vegetables;

b) oxidation of phenolic substances in plant tissues by phenolase (leading to browning);

c) sugar - starch conversion in plant tissues by amylases;

d) post-harvest demethylation of pectic substances in plant tissues (leading to softening of plant tissues during ripening, and firming of plant tissues during processing).

The major factors useful in controlling enzyme activity are: temperature, water activity, pH, chemicals which can inhibit enzyme action, alteration of substrates, alteration of products and pre-processing control.

Chemical Changes

Sensory Quality

The two major chemical changes which occur during the processing and storage of foods and lead to a deterioration in sensory quality are lipid oxidation and non-enzymatic browning. Chemical reactions are also responsible for changes in the colour and flavour of foods during processing and storage. Lipid oxidation rate and course of reaction is influenced by light, local oxygen concentration, high temperature, the presence of catalysts (generally transition metals such as iron and copper) and water activity. Control of these factors can significantly reduce the extent of lipid oxidation in foods.

Non-enzymic browning is one of the major causes of deterioration which occurs during storage of dried and concentrated foods. The non-enzymic browning, or Maillard reaction, can be divided into three stages: a) early Maillard reactions which are chemically well-defined steps without browning; b) advanced Maillard reactions which lead to the formation of volatile or soluble substances; and c) final Maillard reactions leading to insoluble brown polymers.

Colour Changes

Chlorophylls. Almost any type of food processing or storage causes some deterioration of the chlorophyll pigments. Phenophytinisation (with consequent formation of a dull olivebrown phenophytin) is the major change; this reaction is accelerated by heat and is acid catalysed. Other reactions are also possible. For example, dehydrated products such as green peas and beans packed in clear glass containers undergo photo-oxidation and loss of desirable colour.

Anthocyanins. These are a group of more than 150 reddish water-soluble pigments that are very widespread in the plant kingdom. The rate of anthocyanin destruction is pH dependent, being greater at higher pH values. Of interest from a packaging point of view is the ability of some anthocyanins to form complexes with metals such as Al, Fe, Cu and Sn. These complexes generally result in a change in the colour of the pigment (for example, red sour cherries react with tin to form a purple complex) and are therefore undesirable. Since metal packaging materials such as cans could be sources of these metals, they are usually coated with special organic linings to avoid these undesirable reactions.

Carotenoids. The carotenoids are a group of mainly lipid soluble compounds responsible for many of the yellow and red colours of plant and animal products. The main cause of carotenoid degradation in foods is oxidation. The mechanism of oxidation in processed foods is complex and depends on many factors. The pigments may auto-oxidise by reaction with atmospheric oxygen at rates dependent on light, heat and the presence of pro- and antioxidants.

Flavour Changes

In fruit and vegetables, enzymically generated compounds derived from long-chain fatty acids play an extremely important role in the formation of characteristic flavours. In addition, these types of reactions can lead to significant off-flavours. Enzyme-induced oxidative breakdown of unsaturated fatty acids occurs extensively in plant tissues and this yield characteristic aromas associated with some ripening fruits and disrupted tissues.

The permeability of packaging materials is of importance in retaining desirable volatile components within packages, or in permitting undesirable components to permeate through the package from the ambient atmosphere.

Nutritional Quality

The four major factors which affect nutrient degradation and can be controlled to varying extents by packaging are light, oxygen concentration, temperature and water activity. However, because of the diverse nature of the various nutrients as well as the chemical heterogeneity within each class of compounds and the complex interactions of the above variables, generalizations about nutrient degradation in foods will inevitably be broad ones. Vitamins. Ascorbic acid is the most sensitive vitamin in foods, its stability varying markedly as a function of environmental conditions such as pH and the concentration of trace metal ions and oxygen.

The nature of the packaging material can significantly affect the stability of ascorbic acid in foods. The effectiveness of the material as a barrier to moisture and oxygen as well as the chemical nature of the surface exposed to the food are important factors.

For example, problems of ascorbic acid instability in aseptically packaged fruit juices have been encountered because of oxygen permeability of the package and the oxygen dependence of the ascorbic acid degradation reaction. Also, because of the preferential oxidation of metallic tin, citrus juices packaged in cans with a tin contact surface exhibit greater stability of ascorbic acid than those in enamelled cans or glass containers. The aerobic and anaerobic degradation reactions of ascorbic acid in reduced-moisture foods have been shown to be highly sensitive to water activity, the reaction rate increasing in an exponential fashion over the water activity range of 0.1-0.8.

Physical Changes

One major undesirable physical change in food powders is the absorption of moisture as a consequence of an inadequate barrier provided by the package; this results in caking. It can occur either as a result of a poor selection of packaging material in the first place, or failure of the package integrity during storage. In general, moisture absorption is associated with increased cohesiveness.

Anti-caking agents are very fine powders of an inert chemical substance that are added to powders with much larger particle size in order to inhibit caking and improve flowability. Studies in onion powders showed that at ambient temperature, caking does not occur at water activities of less than about 0.4.

At higher activities, however, ($aw > 0.45$) the observed time to caking is inversely proportional to water activity, and at these levels anti-caking agents are completely ineffective. It appears that while they reduce inter-particle attraction and interfere with the continuity of liquid bridges, they are unable to cover moisture sorption sites.

Biological Changes

Microbiological

Micro-organisms can make both desirable and undesirable changes to the quality of foods depending on whether or not they are introduced as an essential part of the food preservation process or arise unintentionally and subsequently grow to produce food spoilage. The two major groups of micro-organisms found in foods are bacteria and fungi, the latter consisting of yeasts and moulds. Bacteria are generally the fastest growing, so that in conditions favourable to both, bacteria will usually outgrow fungi.

Foods are frequently classified on the basis of their stability as non-perishable, semi-perishable and perishable. For example, hermetically sealed and heat processed (e.g. canned) foods are generally regarded as non-perishable. However, they may become perishable under certain circumstances when an opportunity for recontamination is afforded following processing. Such an opportunity may arise if the can seams are faulty, or if there is excessive corrosion resulting in internal gas formation and eventual bursting of the can. Spoilage may also take place when the canned food is stored at unusually high temperatures: thermophilic spore-forming bacteria may multiply, causing undesirable changes such as flat sour spoilage.

Low moisture content foods such as dried fruit and vegetables are classified as semi-perishable. Frozen foods, though basically perishable, may be classified as semi-perishable provided that they are properly stored at freezer temperatures.

The majority of foods (e.g. meat and fish, milk, eggs and most fresh fruits and vegetables) are classified as perishable unless they have been processed in some way. Often, the only form of processing which such foods receive is to be packaged and kept under controlled temperature conditions.

The species of micro-organisms which cause the spoilage of particular foods are influenced by two factors: a) the nature of the foods and b) their surroundings. These factors are referred to as intrinsic and extrinsic parameters.

The intrinsic parameters are an inherent part of the food: pH, a_w, nutrient content, antimicrobial constituents and biological structures. The extrinsic parameters of foods are those properties of the storage environment that affect both the foods and their microorganisms. The growth rate of the micro-organisms responsible for spoilage primarily depends on these extrinsic parameters: temperature, relative humidity and gas compositions of the surrounding atmosphere.

The protection of packaged food from contamination or attack by micro-organisms depends on the mechanical integrity of the package (e.g. the absence of breaks and seal imperfections), and on the resistance of the package to penetration by micro-organisms.

Metal cans which are retorted after filling can leak during cooling, admitting any microorganisms which may be present in the cooling water, even when the double seam is of a high quality. This fact is widely known in the canning industry and is the reason for the mandatory chlorination of cannery cooling water. Extensive studies on a variety of plastic films and metal foils have shown that microorganisms (including mounds, yeasts and bacteria) cannot penetrate these materials in the absence of pinholes.

In practice, however, thin sheets of packaging materials such as aluminium and plastic do contain pinholes. There are several safeguards against the passage of micro-organisms through pinholes in films:

- because of surface tension effects, micro-organisms cannot pass through very small pinholes unless the micro-organisms are suspended in solutions containing wetting agents and the pressure outside the package is greater than that within;
- materials of packaging are generally used in thicknesses such that pinholes are very infrequent and small;
- for applications in which package integrity is essential (such as sterilisation of food in pouches), adequate test methods are available to assure freedom from bacterial recontamination.

Macrobiological

Insect Pests : Warm humid environments promote insect growth, although most insects will not breed if the temperature exceeds about 35 C° or falls below 10 C°. Also many insects cannot reproduce satisfactorily unless the moisture content of their food is greater than about 11%. The main categories of foods subject to pest attack are cereal grains and products derived from cereal grains, other seeds used as food (especially legumes), dairy products such as cheese and milk powders, dried fruits, dried and smoked meats and nuts.

As well as their possible health significance, the presence of insects and insect excrete in packaged foods may render products unsaleable, causing considerable economic loss, as well as reduction in nutritional quality, production of off-flavours and acceleration of decay processes due to creation of higher temperatures and moisture levels. Early stages of infestation are often difficult to detect; however, infestation can generally be spotted not only by the presence of the insects themselves but also by the products of their activities such as webbing, clumped-together food particles and holes in packaging materials.

Unless plastic films are laminated with foil or paper, insects are able to penetrate most of them quite easily, the rate of penetration usually being directly related to film thickness.

In general, thicker films are more resistant than thinner films, and oriented films tend to be more effective than cast films. The looseness of the film has also been reported to be an important factor, loose films being more easily penetrated than tightly fitted films.

Generally, the penetration varies depending on the basic resin from which the film is made, on the combination of materials, on the package structure, and of the species and stage of insects involved. The relative resistance to insect penetration of some flexible packaging materials is as follows:

- excellent resistance: polycarbonate; poly-ethylene-terephthalate;
- good resistance: cellulose acetate; polyamide; polyethylene (0.254 mm); polypropylene (biaxially oriented); poly-vinyl-chloride (unplasticised);
- fair resistance: acrylonitrile; poly-tetra-fluoro-ethylene; polyethylene (0.123 mm);
- poor resistance: regenerated cellulose; corrugated paper board; kraft paper; polyethylene (0.0254 - 0.100 mm); paper/foil/ polyethylene laminate pouch; poly-vinylchloride (plasticised).

Some simple methods for obtaining insect resistance of packaging materials are as following:

- select a film and a film thickness that are inherently resistant to insect penetration;
- use shrink film over-wraps to provide an additional barrier;
- seal carton flaps completely.

Rodents

Rats and mice carry disease-producing organisms on their feet and/or in their intestinal tracts and are known to harbour salmonella of serotypes frequently associated with food-borne infections in humans. In addition to the public health consequences of rodent populations in close proximity to humans, these animals also compete intensively with humans for food. Rats and mice gnaw to reach sources of food and drink and to keep their teeth short. Their incisor teeth are so strong that rats have been known to gnaw through lead pipes and unhardened concrete, as well as sacks, wood and flexible packaging materials.

Proper sanitation in food processing and storage areas is the most effective weapon in the fight against rodents, since all packaging materials apart from metal and glass containers can be attacked by rats and mice.

Methods of Reducing Deterioration

A knowledge of deterioration factors and the way they act, including the rates of deterioration to a specific category of food, means that it is possible to list the ways of lowering or stopping the action and obtaining fruit and vegetable preservation.

In order to maintain their nutritional value and organoleptic properties and because of technical-economical considerations, not all the identified means against deterioration actually have practical applications for fruit and vegetable preservation.

Technical Methods of Reducing Food Deterioration

These technical means can be summarised as follows:

Physical	Heating
	Cooling
	Lowering of water content Drying/dehydration. Concentration
	Sterilising filtration
	Irradiation
	Other physical means (high pressure, vacuum, inert gases)
Chemical	Salting
	Smoking
	Sugar addition
	Artificial acidification
	Ethyl alcohol addition
	Antiseptic substance action
Biochemical	Lactic fermentation (natural acidification)
	Alcoholic fermentation

This classification of methods of reducing deterioration presents some difficulties because their preservation effects are physical, physico-chemical, chemical and biochemical complex phenomena which rarely act in isolation. Normally they take place together or one after the other. From the whole list of possible methods of reducing deterioration, over the years, some procedures for fruit and vegetable preservation have found practical application.

Table: *Procedures for Fruit and Vegetable Preservation*

Procedures	Practical applications
Fresh storage	Fruits, vegetables
Cold storage	Fruits, vegetables
Freezing	Fruits, vegetables
Drying/dehydration	Fruits, vegetables
Concentration	Fruit and vegetable juices

Contd...

Chemical preservation	Fruit semi-processed
Preservation with sugar	Fruit products/preserves
Pasteurization	Fruit and vegetable juices
Sterilisation	Fruits, vegetables
Sterilising filtration	Fruit juices
Irradiation	Fruits, vegetables

These preservation procedures have two main characteristics as far as being applied to all food products is concerned:

- some of them are applied only to one or some categories of foods; others can be used across the board and thus a wider application (cold storage, freezing, drying/dehydration, sterilisation, etc.);
- some guarantee food preservation on their own while others require combination with other procedures, either as principal or as auxiliary processes in order to assure preservation (for example smoking has to be preceded by salting).

Combined Preservation Procedures

In practice preservation procedures aim at avoiding microbiological and biochemical deterioration which are the principal forms of deterioration. Even with all recent progress achieved in this field, no single one of these technological procedures applied alone can be considered wholly satisfactory from a microbiological, physico-chemical and organoleptic point of view, even if to a great extent the food value is assured. Thus, heat sterilisation cannot be applied in order to destroy all micro-organisms present in foods without inducing non desirable modifications. Preservation by dehydration/drying assures microbiological stability but has the drawback of undesirable modifications that appear during storage: vitamin losses, oxidation phenomena, etc.

Starting with these considerations, the actual tendency in food preservation is to study the application of combined preservation procedures, aiming at the realisation of maximum efficiency from a microbiological and biological point of view, with reduction to a minimum of organoleptical degradation and decrease in food value.

The principles of combined preservation procedures are:

- avoid or reduce secondary (undesirable) effects in efficient procedures for microbiological preservation;

- avoid qualitative degradation appearing during storage of products preserved by efficient procedures from a microbiological point of view;
- increase microbiological efficiency of preservation procedures by supplementary means;
- combine preservation procedures in order to obtain maximum efficiency from a microbiological point of view, by specific action on various types of micro-organisms present;
- establish combined factors that act simultaneously on bacterial cells.

Research and applications in this direction were followed by microbiological and biochemical way, obtaining a serial of combination of preservation procedures with the possibility of application in industrial practice. [unclear]

Fresh fruit and vegetable storage can be combined with:

- storage in controlled atmosphere where carbon dioxide and oxygen levels are monitored, increasing concentration of CO2 and lowering that of oxygen according to fruit species. Excellent results were obtained for pomace fruit; in particular the storage period for apples has been extended. Application of this combined procedure requires airtight storage rooms.
- storage in an environment containing ethylene oxide; this accelerates ripening in some fruit: tomatoes, bananas, mangoes, etc.

Cold storage can be combined with storage in an environment with added of carbon dioxide, sulphur dioxide, etc. according to the nature of product to be preserved.

Preservation by drying/dehydration can be combined with: freezing: fresh fruit and vegetables are dehydrated up to the point where their weight is reduced by 50% and then they are preserved by freezing.

This procedure (freeze-drying) combines the advantages of drying (reduction of volume and weight) with those of freezing (maintaining vitamins and to a large extent organoleptic properties).

A significant advantage of this process is the short drying time in so far as it is not necessary to go beyond the inflexion point of the drying curve. The finished products after defreezing and rehydration/ reconstitution are of a better quality compared with products obtained by dehydration alone.

- cold storage of dried/dehydrated vegetables in order to maintain vitamin C; storage temperature can be varied with storage time and can be at -8° C for a storage time of more than one year, with a relative humidity of 70-75 %.
- packaging under vacuum or in inert gases in order to avoid action of atmospheric oxygen;, mainly for products containing beta-carotene.
- chemical preservation: a process used intensively for prunes and which has commercial applications is to rehydrate the dried product up to 35 % using a bath containing hot 2 % potassium sorbate solution. Another possible application of this combined procedure is the initial dehydration up to 35% moisture followed by immersion in same bath as explained above; this has the advantage of reducing drying time and producing minimum qualitative degradation. Both applications suppress the dehydrated products reconstitution (rehydration) step before consumption.
- packaging in the presence of desiccants (calcium oxide, anhydrous calcium chloride, etc.) in order to reduce water vapour content in the package, especially for powdered products.

Preservation by concentration, carried out by evaporation, is combined with cold storage during warm season for tomato paste (when water content cannot be reduced under the limit needed to inhibit moulds and yeasts, e.g. a_w = 0.70...0.75).

Chemical preservation is combined with:

- acidification of food medium (lowering pH);
- using combined chemical preservatives.

Preservation by lactic fermentation (natural acidification) can be combined with cold storage for pickles in order to prolong storage time or shelf-life. Preservation with sugar is combined with pasteurization for some preserves having a sugar content below 65%.

Bibliography

Adams, H.P. and R.B. Bushnell.: *Dry Cow Feeding and Management*, A Western Regional Extension Publication, 1979.

Ashraf, Syed Mahboob: *Handbook of Fruit and Vegetable Products*, Agrobios, Delhi, 2008.

Bajpai, Amita: *Recent Trends in Animal Nutrition and Feed Technology for Livestock, Pets and Laboratory Animals*, International, 2003.

Boekel, M. A. J. S. van: *Dairy Technology: Principles of Milk Properties and Processes,* New York: M. Dekker, 1999.

Chakraborty, Sudip: *Food Security and Child Labour : The Case of a Hazardous Occupation*, Deep and Deep, Delhi, 2011.

Copeland, R. A., *Enzymes: A Practical Introduction to Structure,Mechanism, and Data Analysis*, New York, 1999.

Deepti, R. Haritash and Tej Verma: *Promoting Community Nutrition Through Participatory Monitoring and Evaluation*, Advance, 2001.

Dharamvir Hota: *Modern Biotechnology in Plant Breeding*, Gene Tech Books, Delhi, 2007.

Fersht, A.: *Enzyme Structure and Mechanism*, W. H. Freeman, New York, 1977.

Gautam, U.S. and Ram Chand: *Decision-Making of Dairy Beneficiaries: Role of Aspiration, Motivation and Knowledge*, Om Pub, Delhi, 2010.

Gutfreund, H. : *Enzymes: Physical Principles*, Wiley-Interscience, London, 1972.

Hemantaranjan, A.: *Advancements in Iron Nutrition Research*, Scientific, Delhi, 1995.

Heslot, H. and Gaillardin, C.: *Molecular Biology and Genetic Engineering of Yeast*, CRC Press, Boca Raton, FL 1992.

Jacobson, M. : *Safe Food: Eating Wisely in a Risky World,* Washington, DC: Living Planet Press, 1991.

Jha, S N: *Dairy and Food Processing Plant Maintenance : Theory and Practice,* International Book Distributing, Delhi, 2006.

Krimsky, Sheldon : *Biotechnics and Society: The Rise of Industrial Genetics*, New York: Praeger Publishers, 1991.

Law, Barry A.: *Microbiology and Biochemistry of Cheese and Fermented Milk*, London: Blackie Academic & Professional, 1997.

Lois D. McBean.: *Handbook of Dairy Foods and Nutrition*, Boca Raton, Fla.: CRC Press, 1999.

Mayrose, V.B.: *Handbook of livestock Management Techniques*, Macmillan Publishing Co., New York, 1981.

Mowery, D. C.: *Technology and Wealth of Nations*, Stanford: Stanford University Press, 1992.

Nyholt, D.H. : *The Vitamin & Herb Guide,* Alberta, Canada: Global Health Ltd., 1992.

Oldham P.: *Cost of Production of Major Tree Crops in Dominica*, Roseau, Ministry of Agriculture, 1991.

Parmar, B.S.: *Pesticides, Crop Protection and Environment*, Oxford & IBH, Delhi, 1995.

Piszkiewicz, D.: *Kinetics of Chemical and Enzyme-Catalyzed Reactions*, Oxford University Press, New York, 1977.

Ramakant Sharma: *Chemical and Microbiological Analysis of Milk and Milk Products,* International Book Distributing, 2006.

Rhykerd Charles L.: *The Cycles of Plant and Animal Nutrition*, Scientific American Books, San Francisco 1976.

Sharma, Ramakant : *Chemical and Microbiological Analysis of Milk and Milk Products,* International Book Distributing, 2006.

Sjaastad, V.: *Physiology of Domestic Animals*, International Book Distributing Co., Delhi, 2005.

Thackray, Arnold : *Private Science: Biotechnology and the Rise of the Molecular Sciences,* Philadelphia: University of Pennsylvania Press, 1998.

Tomar, S.K. and Gunjan Goel: *Applied Dairy and Food Microbiology*, Agrotech, 2005.

Upadhyay, K.G. and Vyas, S.H.: *Composition of Camel's Milk*, Gujarat Agric. University, 1982.

Venkateshwara Rao: *Dairy Farm Business Management*, Biotech Books, Delhi, 2008.

Wade, C. : *Helping Your Health with Enzymes,* New York, U.S.A: Universal-Award House, 1971.

Walsh, C. T. : *Enzymatic Reaction Mechanisms*, W.H. Freeman, New York, 1979.

Yadav, D.S.: *Advances in Vegetable Production*, Researchco Book Centre, Delhi, 2005.

Index

❑❑❑